FOREWORD

Understanding the factors that influence public perception and confidence in the area of radioactive waste management is of strategic interest to the Radioactive Waste Management Committee (RWMC) of the Nuclear Energy Agency (NEA). The Forum on Stakeholder Confidence (FSC) is a working party of the RWMC and acts as the centre for the exchange of opinions and experiences across institutional and non-institutional boundaries. In order to distil the lessons that can be learnt, the FSC promotes open discussion across the entire spectrum of stakeholders in an atmosphere of trust and mutual respect. Workshops in national contexts have been identified as the preferred means for interaction.

On 18 May 2001, the Finnish Parliament ratified the Decision in Principle on the final disposal facility for spent nuclear fuel in Olkiluoto, Eurajoki. The Government had made a positive decision earlier, at the end of 2000, and in compliance with the Nuclear Energy Act, the Parliament's ratification was then required. The decision is valid for the spent fuel generated by the existing Finnish nuclear power plants and means that the construction of the final disposal facility is considered to be in line with the overall good of society. Earlier steps included, amongst others, the approval by the nuclear regulatory body and the host community. Future steps include the construction of an underground rock characterisation facility, ONKALO (2003-2004), and applications for separate construction and operating licences for the final disposal facility (from about 2010).

How did this political and societal decision come about? The workshop provided the opportunity to present the history leading up to the Decision in Principle, and to examine future perspectives with an emphasis on stakeholder involvement.

By gathering Finnish stakeholders, those who expressed favour and opposition, as well as observer-participants from other countries, a joint reflection on a complex situation could be undertaken. The workshop provided a review of the Finnish programme, by and for the FSC, as well as by and for the Finnish stakeholders.

The NEA Secretariat wishes to thank Posiva Oy, VTT Energy, the Finnish Radiation and Nuclear Safety Authority (STUK), and the Finnish Ministry of Trade and Industry (Energy Department) for their collaboration in organising and hosting the workshop.

Stepwise Decision Making in Finland for the Disposal of Spent Nuclear Fuel

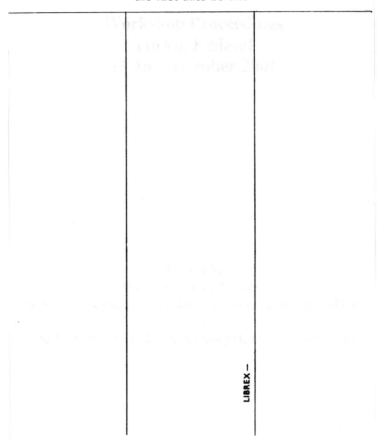

Books are to be returned on or before
the last date below.

LIBREX —

D
621.4838
STE

NUCLEAR ENERGY AGENCY
ORGANISATION FOR ECONOMIC CO-OPERATION AND DEVELOPMENT

ORGANISATION FOR ECONOMIC CO-OPERATION AND DEVELOPMENT

Pursuant to Article 1 of the Convention signed in Paris on 14th December 1960, and which came into force on 30th September 1961, the Organisation for Economic Co-operation and Development (OECD) shall promote policies designed:

- to achieve the highest sustainable economic growth and employment and a rising standard of living in Member countries, while maintaining financial stability, and thus to contribute to the development of the world economy;
- to contribute to sound economic expansion in Member as well as non-member countries in the process of economic development; and
- to contribute to the expansion of world trade on a multilateral, non-discriminatory basis in accordance with international obligations.

The original Member countries of the OECD are Austria, Belgium, Canada, Denmark, France, Germany, Greece, Iceland, Ireland, Italy, Luxembourg, the Netherlands, Norway, Portugal, Spain, Sweden, Switzerland, Turkey, the United Kingdom and the United States. The following countries became Members subsequently through accession at the dates indicated hereafter: Japan (28th April 1964), Finland (28th January 1969), Australia (7th June 1971), New Zealand (29th May 1973), Mexico (18th May 1994), the Czech Republic (21st December 1995), Hungary (7th May 1996), Poland (22nd November 1996), Korea (12th December 1996) and the Slovak Republic (14 December 2000). The Commission of the European Communities takes part in the work of the OECD (Article 13 of the OECD Convention).

NUCLEAR ENERGY AGENCY

The OECD Nuclear Energy Agency (NEA) was established on 1st February 1958 under the name of the OEEC European Nuclear Energy Agency. It received its present designation on 20th April 1972, when Japan became its first non-European full Member. NEA membership today consists of 28 OECD Member countries: Australia, Austria, Belgium, Canada, Czech Republic, Denmark, Finland, France, Germany, Greece, Hungary, Iceland, Ireland, Italy, Japan, Luxembourg, Mexico, the Netherlands, Norway, Portugal, Republic of Korea, Slovak Republic, Spain, Sweden, Switzerland, Turkey, the United Kingdom and the United States. The Commission of the European Communities also takes part in the work of the Agency.

The mission of the NEA is:

- to assist its Member countries in maintaining and further developing, through international co-operation, the scientific, technological and legal bases required for a safe, environmentally friendly and economical use of nuclear energy for peaceful purposes, as well as
- to provide authoritative assessments and to forge common understandings on key issues, as input to government decisions on nuclear energy policy and to broader OECD policy analyses in areas such as energy and sustainable development.

Specific areas of competence of the NEA include safety and regulation of nuclear activities, radioactive waste management, radiological protection, nuclear science, economic and technical analyses of the nuclear fuel cycle, nuclear law and liability, and public information. The NEA Data Bank provides nuclear data and computer program services for participating countries.

In these and related tasks, the NEA works in close collaboration with the International Atomic Energy Agency in Vienna, with which it has a Co-operation Agreement, as well as with other international organisations in the nuclear field.

TABLE OF CONTENTS

EXECUTIVE SUMMARY

1. Background

On 18 May 2001, the Finnish Parliament ratified the Decision in Principle on the final disposal facility for spent nuclear fuel at Olkiluoto, within the municipality of Eurajoki. The Municipality Council and the government had made positive decisions earlier, at the end of 2000, and in compliance with the *Nuclear Energy Act*, Parliament's ratification was then required. The decision is valid for the spent fuel generated by the existing Finnish nuclear power plants and means that the construction of the final disposal facility is considered to be in line with the overall good of society. Earlier steps included, amongst others, the approval of the technical project by the Safety Authority. Future steps include construction of an underground rock characterisation facility, ONKALO (2003-2004), and application for separate construction and operating licences for the final disposal facility (from about 2010).

How did this political and societal decision come about? The FSC Workshop provided the opportunity to present the history leading up to the Decision in Principle (DiP), and to examine future perspectives with an emphasis on stakeholder involvement.

This Executive Summary gives an overview of the presentations and discussions that took place at the workshop.[1] It presents, for the most part, a factual account of the individual presentations and of the discussions that took place. It relies importantly on the notes that were taken at the meeting. Most materials are elaborated upon in a fuller way in the texts that the various speakers and session moderators contributed for these proceedings. The structure of the Executive Summary follows the structure of the workshop itself.

Complementary to this Executive Summary and also provided with this document, is a NEA Secretariat's perspective aiming to place the results of all discussions, feedback and site visit into an international perspective.

2. Introduction to the workshop

Carol Kessler, Deputy Director General of the OECD Nuclear Energy Agency (NEA), welcomed the participants of the second FSC workshop. Ms. Kessler emphasised the importance of collaboration between the technical community and civil society in implementing technologies, which have apparent negative aspects, such as nuclear energy. She pointed out that the OECD has recently been engaged in supporting such collaboration. Important steps in this development included the 1999 and the 2000 Ministerial Council meetings, when the Ministers recognised the responsibility to ensure transparency and clarity in policy making and asked the OECD to assist governments in collaborating with civil society organisations (CSO). In 2000, the OECD established its Forum 2000 to seek CSO

---

1. The programme of the Workshop is provided in Annex 1.

input to the Ministerial Agenda and held one again in 2001. The Public Management Service (PUMA) of OECD has been established to assist governments in collaborating with civil society.

Ms. Kessler reviewed NEA activities in this field, including a recent workshop held by the Committee on Radiation Protection and Public Health on "Better Integration of Radiation Protection in Modern Society", and a workshop organised by the Committee on Nuclear Regulatory Activities entitled "Investing in Trust – Nuclear Regulators and the Public". She noted that OECD guidance in this field should remain limited, since so much of the government/civil society dialogue is nation specific. On the other hand, Ms. Kessler acknowledged the importance of exchanging ideas in this field and of the role of NEA in assisting governments to learn from each other. The Turku Workshop is an excellent opportunity for discussing what aspects of the Finnish process may be unique for Finland and what may be more generally applicable.

Yves Le Bars, Chairman of the FSC, summarised the objectives and the organisation of the workshop. By gathering key Finnish stakeholders and observers/participants from the FSC community, the workshop provides a review of the Finnish process by and for both groups.

The workshop was planned to focus on three issues including *the stepwise decision-making process, stakeholder involvement,* and *confidence building.* Mr. Le Bars introduced the highly interactive format, which was chosen for the meeting. Presentations by *plenary speakers* were to be followed by: (1) roundtable discussions facilitated by *facilitators*; (2) plenary sessions reviewing the results of these discussions, directed by *session moderators*; and (3) cross-cutting overview reports by *thematic rapporteurs* compiled on the basis of their observations at the workshop.

3.   Session I: Background to the Decision in Principle

Ilkka Ruostetsaari from the University of Tampere spoke of the possible reasons for the apparent contradiction between Parliament's consensus on the construction of the repository and the ongoing debate on the future of nuclear power in Finland. He identified two kinds of explanation, the first one residing in the characteristics of Finnish political culture, and the second in some contextual factors.

Characteristics of Finnish political culture, which he found significant, include the Nordic concept of autonomy, the Lutheran religion, capitalism, nationalism, and the power of the ruler and the bureaucracy. There is "consensual" rather than "adversarial" policy-making, so much so that parties that win elections do not necessarily get to govern. Main contextual factors which facilitated agreement include the legal framework, the political situation, the actors' adaptability, the competition between two municipalities for hosting the facility, the weakness of opposition movements, the non-partisan role of the media, and the fact that the debate on the waste disposal issue was de-linked from the political debate on nuclear power. Finally, Finns display great trust in technology and education, and the respect for institutions and the government is very high.

Jussi Manninen from the Ministry of Trade and Industry briefly presented the history of nuclear energy production and waste management in Finland. He listed a series of changes in the legislative framework, which led to the current Parliament ratification of the DiP related to the final disposal facility ("facility DiP"). Of special importance was the 1983 government Decision in Principle ("strategic DiP"), which excluded storage as a long-term method for waste management and required that a site for final repository be selected by the year 2000. Another important step was the 1987 Nuclear Energy Act, which defined the responsibilities of various actors in nuclear waste management, and the 1994 amendment of this Act requiring the final disposal of any spent fuel produced in Finland to take place in Finland.

In addition, Mr. Manninen indicated several issues concerning the legal background. For example, some provisions of the 1994 Act on Environmental Impact Assessment overlap with the Nuclear Energy Act.

Jukka Laaksonen from STUK indicated that the regulator has played an important role in Finland since the very beginning of the nuclear era. He acknowledged the multidimensional challenges faced in order to reach the DiP safety issues, societal issues, and the timing and scheduling of the decision-making process. He pointed out that in the facility DiP stage, no definite conclusion on the safety of the proposed disposal concept was required. Only a preliminary safety appraisal was needed, stating that nothing had been found which would raise doubts about the potential to achieve the required safety level.

The facility DiP means that the disposal site has been chosen, and there is a firm commitment by a municipality to host the facility. It also means that safety studies can be continued at the site. According to Mr. Laaksonen, at least ten more years will be needed to receive sufficient assurance on safety, so that a construction permit for the repository can be issued.

4.   Session II: Process of Stepwise Decision Making in Finland

Veijo Ryhänen from Posiva Oy reviewed the obligations of the operator in having to take both technical and social steps and gave a summary of the past and future milestones, and the key activities of the siting process. The obligation to take social steps come from the fact that the DiP decision is stipulated on "the overall good of society" and that the local community has veto power. The Environmental Impact Assessment (EIA) process was especially useful for reaching out to the public and learning from it. The discussion of non-implementation of a solution (the zero alternative) versus implementing a solution was essential for a shared understanding of the necessity to move forward. He suggested that one item that contributed to the confidence underlying the DiP is the fact that both at Olkiluoto and Lovisa disposal facilities for LLW and ILW and interim storage facilities for spent fuel are operational with an excellent record. He emphasised the importance of having a stepwise process relying on a clear legal background and a long-term commitment from the part of the government.

Tero Varjoranta from STUK outlined the risk communication programme conducted by STUK in potential host communities. The regulator's public information programme addressing the concerns, expectations, and information needs of local residents proved to be successful. This also raised the profile of STUK. People wanted STUK to have a referee role and to be on the side of the municipality in what regards health concerns. Indeed, STUK has implemented an internal mandate of providing "the best information available" to the public and will go the municipality when asked. STUK also tries to assist local media, which may lack adequate resources for researching articles. Mr. Varjoranta commented that it is not obvious that the methods that are effective in Finland could be readily transferable to other countries.

Kimmo Tuikka, representative of the Kivetty Movement, spoke of the protest movement that emerged in the town of Äänekoski when a candidate site was identified in its Kivetty area. In particular, Äänekoski area had long suffered of an image problem in Finland and the disposal facility would have re-opened an issue that is now solved. Safety of the transport of the waste to Äänekoski was questioned, the lack of infrastructure was an issue as well and, overall, the movement came to the conclusion that that there would be no net benefit to the community from hosting a disposal facility of spent fuel. The Kivetty Movement entered in the political arena and was able to have a new city council elected that was more sceptical than the previous one about hosting the disposal facility. Mr. Tuikka recognised the wealth of information that was provided during the EIA process, but expressed doubts about the independence of the EIA study from the influence of Posiva Oy. He also commented on the fact that the regulator was not so visible as it now is, which is a positive development. He questioned the separation of the nuclear waste management issue from that of the future of nuclear power and whether Finland may end up receiving waste from other EU countries.

Altti Lucander, member of Eurajoki Municipality Council, analysed the facility siting process from a local perspective. He indicated that key roles in the acceptance of the facility by the residents of Eurajoki were: the expected economic and social benefits, moral responsibility for the management of waste, confidence in TVO which runs the Olkiluoto NPP in the Eurajoki municipality, and a high level of trust in the safety authorities. He observed that the process envisaged by the State to come to a decision was easy to work with, and that the responsibility for acceptance of a facility lies foremost with the community who must rely on the regulator as their specialist advisor.

The **roundtable discussions** (session moderator: **Michael Aebersold**) resulted in the following main conclusions:

a) *What were the most important steps in the decision-making process for the different stakeholders?*

- The 1983 government decision specifying the milestones for final disposal.

- The 1994 government decision banning the export and import of nuclear waste.

- The Environmental Impact Assessment (EIA) process.

- Safety review by the regulatory body (STUK) and government Decision in Principle (DiP) of 2000.

- Approval of the DiP by Parliament in 2001.

b) *What influenced the process and the outcome?*

- The institutional framework (DiP, EIA and STUK), the step-by-step decision process, the simple organisational structure, the political decision to prohibit the export of fuel and the early introduction of the concept of geological disposal.

- Broad political consensus, on a national and regional level, regarding the site of the disposal facility. High level of public confidence in the host community, and competition among potential host communities.

- Participation of stakeholders and transparency of the process.

c) *What are the lessons learnt?*

- It is important to recognise that a problem exists, which needs to be solved and can be solved.

- Confidence and trust in the regulatory body and the implementors is crucial.

- The municipality is a major stakeholder, and its veto right is a very important element.

- Differences between risk perception by experts and lay people have to be understood and public concerns need to be taken into account.

- The following elements were key factors of success:

    - DiP as part of a stepwise procedure and as principal decision for implementation.

    - EIA as a structure and guide for public involvement and participation.

    - STUK as a regulatory body which creates confidence.

5. Session III: Stakeholder Involvement, Particularly in the Environmental Impact Assessment

Anne Väätäinen from the Ministry of Trade and Industry addressed the role of the Ministry in the public participation process. Two public hearings were organised by the Ministry in connection with the EIA process (prescribed by the *EIA Act*) and the government DiP (specified by the *Nuclear Energy Act*), respectively. Although she evaluated the hearings as successful events of public participation, she also indicated some deficiencies, which she attributed to the lack of harmonisation between the *EIA Act* and the *Nuclear Energy Act*. The EIA process highlighted the need to discuss alternatives and their impacts and, in particular, the "zero option".

Pekka Hokkanen from the Tampere University acknowledged the relevance of the EIA as a most important instrument for supporting a political decision in this case and noted that the EIA legislation underscores public involvement. He described participation as negligible, however, and decreasing throughout the process. He attributed the low level of participation to the lack of such participatory traditions, the lack of familiarity with this instrument, the lack of confidence in the effectiveness of participation, the tiredness and exhaustion of some stakeholders, and the uneven distribution of the resources amongst stakeholders.

Jorma Jantunen from Uusimaa Regional Environmental Centre addressed the role of Finnish environmental authorities in the EIA process. The regional authority focusing on local environmental impacts was largely satisfied with the EIA, in particular with the information it provided. The Ministry for Environment, which overviews that the legal requirements are met, was satisfied with the level of participation but was more critical about a number of items (e.g., Posiva's advertising campaign). The Finnish Environment Institute raised concerns about the "goal-directedness" of the process: it observed that while the scope of the EIA was local, it had national dimensions and wider participation than just local was appropriate. They would also have wished to see more information on alternatives. The EIA process took into account also international conventions. Namely, the neighbouring countries were informed, were able to provide comments, and a positive statement was obtained from their part.

Juhani Vira of Posiva Oy indicated that the EIA was an important initiative for stakeholder involvement and of information for the DiP. He reviewed the many stakeholder voices that took part in the EIA process, recalled that the final report was issued in three languages, and that all views were recorded including the dissenting ones. The number of participants may have been limited, vis-à-vis expectations, but the quality of the debate was high. On specific points, the following was learnt: people questioned whether experts can claim knowledge in long-term safety. The evocation of the image of the community may be also a "cover" for something that people are uneasy with; the discussion of alternatives is very important for the political decision that will follow; retrievability arose as an issue and firmed up as a positive feature if provisions for retrievability are implemented. Indeed, one important result of the process was the government decision that requires that the spent fuel must be retrievable even after closing the disposal facility. An area of concern was that of social impacts within the community, which was addressed through a comprehensive social impact assessment study.

Thomas Rosenberg, representative of the Lovisa movement, strongly criticised the EIA process, which he evaluated as a "long, frustrating, co-optative, and scientifically camouflaged" and serving only for legitimising the decisions. The movement did get radio coverage, and succeeded in changing the agenda of the discussions (e.g., put the alternative method of Dry Rock Deposition on the agenda), raising local awareness about nuclear issues, and preventing the siting in Lovisa. The movement was against disposal, preferring long-term surface storage at Lovisa rather than an underground repository at Eurajoki. The division in the community was exacerbated by current issues in cultural demarcation amongst Finnish and Swedish speakers.

Antti Leskinen from Discurssi Oy emphasised the importance of the scoping phase of the EIA in structuring the investigations according to the needs of the local public. Whilst the EIA processes in Finland do not have necessarily a scoping phase, this one did. He considered this EIA a process of "good quality" vis-à-vis other EIAs in Finland. He acknowledged that the effectiveness of this EIA as a framework for public participation was questioned by several stakeholders in that participation was relatively low. On the other hand, even opponents found that sufficient information was made available, and when people do not come to public meetings it is not necessarily justified to think that they are against.

As a result of **roundtable discussions** (session moderator: **Hideki Sakuma**), the following main conclusions were drawn:

a)   *Was the stakeholder involvement process sufficient?*

- A majority of workshop participants shared the view that the EIA process provided sufficient opportunities for stakeholder participation. The leader of a local protest movement, however, claimed that chances provided for various stakeholders to participate and influence decisions were far from equal.

- The participation of STUK was especially acknowledged by a majority of participants.

b)   *Did you receive all the information you needed for your involvement?*

- A majority of the Finnish participants found that sufficient information was available. Some claimed that there was too much information.

- Some claimed that information provided about alternative waste management methods was insufficient.

- It was mentioned that due to the lack of resources, opponents could not hire independent experts.

c)   ***What are the lessons learnt?***

- It is important that the role of EIA in the siting process, as well as the role of stakeholder involvement in the EIA process be clear from the beginning.

- Stakeholders should be allowed to participate from the very early stages of the siting process.

- Public interest in participation can be maintained only if stakeholders believe that they can have an influence on key decisions.

- Continued dialogue between the implementors and local people is crucial.

d)   ***How could your involvement be improved in the future?***

- The complexity of EIA should be simplified, public participation should be made easier.

- More attention should be paid to informing people.

- More attention should be paid to listening to people and responding to their concerns.

- Resources should be provided for less powerful stakeholders to assure that they have fair chances for effective participation.

6.   Session IV: What Gives Confidence to the Various Categories of Stakeholders?

Janina Andersson from the Green Parliamentary Group acknowledged the open seminars where everybody had the possibility to talk. Indeed NGOs had the chance to express themselves also in Parliament and, overall, the openness of Posiva is commendable. The "commercial" attitude of Posiva in their campaign was less appreciated though. A large component of the positive decision by Parliament was that the problem cannot be passed on to others, and it has to dealt with within the national borders while the know-how is available. In particular, the Green party voted in favour of the DiP also on the consideration that it felt it had an obligation to find a national solution to a problem that was accentuated when it voted, in 1994, for not exporting out of the country any wastes produced in Finland, including spent fuel. The DiP preserves the good of the community right now, it may be up to the community to re-decide 100 years from now. She observed that the process is not over yet, and it is important not to speed it up needlessly. Some questions mentioned are: the optimal depth of the repository, salinity and temperature effects at depth, and the properties of the bentonitic materials on which performance seems to rely importantly. Parliament will want to consider all the facts before making a final decision on constructing a repository.

Altti Lucander, member of Eurajoki Municipality Council, pointed out that the high level of confidence existing at Eurajoki can be attributed to: the excellent safety record of the nuclear power plant of Olkiluoto; the openness of both the implementor and the regulator responsible for spent fuel management, in particular the information given is of good quality, transparent, and is provided quickly to the community; many meetings on topical issues, including those facilitated by the EIA as a platform for dialogue; the fact that a number of inhabitants of Eurajoki work at the NPP; and the existence of local liaison groups facilitating the dialogue between the municipality and TVO.

In the past the Community had voted for the principle of not accepting disposal of nuclear waste in its territory. The community turned around after both the export and import of spent fuel were prohibited by law in 1994. Other milestones important for the community positive decision were a study on the economical competitiveness of the municipality in 1997, and a 1998 analysis known as the "Olkiluoto vision". Eurajoki has also been careful to maintain dialogue with neighbouring communities.

Tapio Litmanen from the University of Jyväskylä analysed the role of social science for the national waste management programme. Social science was integrated in the decision-making process in the mid-1990s. Two forces acted in that direction: the veto power given to the community by the *Nuclear Energy Act* and the preparation of the EIA process. A number of studies were performed with public funding that helped identify questions for the EIA to address, how to implement and how to evaluate the EIA, etc. At present, a project is ongoing at the University of Tampere to evaluate what lessons are to be learnt on, and from, the EIA process. He argued that applied social science research, as well as theoretical research could be very helpful to improve understanding the nature and roots of controversies, and finding ways to increase mutual trust.

The paper by **Seppo Vuori** and **Kari Rasilainen** from VTT Energy gave an introduction on the Public Sector's Nuclear Waste Management Research Programme aimed at investigating both technical and social science issues related to the spent fuel management. The research programme, which was independent of Posiva's own research, has been underway since 1989 and has supported the activities of the authorities. The technical programme has increased confidence in numerous technical areas, including illustrations of radiological impacts and of alternative management routes in Finland. From their R&D perspective, the authors identified five points that were instrumental for the positive outcome of the DiP: clear legislative requirements; veto power of the community; public involvement through the EIA; independent review by the regulator; and the decisive role of Parliament.

Main conclusions emerging from the **roundtable discussions** (session moderator: **Simon Webster**) are as follows:

1.   *What was important for developing confidence? How would you rank the various measures?*

- In general, the fairness and transparency of the decision-making process were emphasised as key factors of trust and acceptance.

- For the municipality, the right of veto, the clear government strategy, and public participation as defined by EIA were most significant.

- Some participants considered institutional measures as most important, followed by the social and technical measures.

- Some emphasised the importance of maintaining the dialogue between various stake-holders throughout the whole duration of the project.

2.   *What were positive and negative experiences for gaining confidence and trust?*

- For positive experiences, see above

- Some parties (e.g., Ministry of Trade and Industry, research organisations) were criticised for not being neutral or sufficiently competent.

- Some tools (e.g., Posiva's information campaign, public surveys) were criticised as unfair or inappropriate.

- Concerns were expressed over the past changes of policy regarding the export of waste, the lack of control by Parliament after approving following the DiP, and Finland's being the first country to establish a repository.

3.   *What are the lessons learnt? What should be done to improve confidence and trust?*

- Openness, honesty, early and continuous participation of a variety of stakeholders are key factors.

- Adopting a step-wise approach with public outreach increases the chances of success.

- The process is not over yet, the dialogue needs to be continued.

- The lessons learnt from the Finnish process are only partially transferable to other countries.

7.   Session V: Conclusions, Assessment and Feedback

In this session, reports from four thematic rapporteurs observing the workshop were presented.

Frédéric Bouder from OECD/PUMA analysed the public governance aspects of the Finnish case. He found the adaptability of the siting process, especially the progressive normalisation of stakeholder involvement, of high significance. Another key feature he identified, was the combination of municipality vote with a final decision by the National Parliament.

Mr. Bouder suggested that stakeholder involvement be carefully planned and the following questions be given serious consideration: Who is involved at what stage? Who monitors the agenda? What tools are used to inform/consult/participate? How are messages translated? What feedback is given? How open and transparent is the process? Finally, based on the findings of the workshop and previous research studies, Mr. Bouder suggested that stakeholder involvement in the nuclear field meet a set of, cross-culturally relevant, criteria (see Outlook Remarks).

The report by **Claire Mays** from Institut Symlog (France) took a social-psychological perspective. The central concept she chose for her analysis was the contrast between "in-groups" and "out-groups", i.e., those who have a dominant position in defining a situation and those who are in a position of exclusion. By analysing the jokes offered and the vocabulary used by workshop participants, she illustrated the difficulty to reach and include in social dialogues individuals from out-groups (e.g., members of opposition movements, "ordinary people"), and the tendency for patronising them.

By reacting to certain deficiencies of public participation in the Finnish case, Ms. Mays suggested that other modes of expression, besides written contributions be applied, in which stakeholders could have confidence that they would affect decision making. She emphasised that in order to increase participation, the expression of unsystematic knowledge, beliefs, values, preferences, and feelings should also be encouraged.

Finally, Ms. Mays pointed out that although at the workshop the uniqueness of Finnish culture was frequently mentioned as a key factor of successful siting, the relative influence of culture and other features of individual, organisational, political, social, historical contexts could not be determined on the basis of available data.

Anna Vári, from the Hungarian Academy of Sciences, presented her observations about community development and siting issues. According to her analysis, for the majority of Eurajoki residents, the balance of anticipated positive and negative impacts of the planned facility was positive, and this was a crucial factor contributing to local acceptance.

The way of sharing the benefits and burdens of nuclear power production raises general questions about fairness[2]. Ms. Vári pointed out that there is no single morally correct way for allocating benefits and burdens between stakeholders, and the history of nuclear waste management policy in Finland reveals the plurality and the changeable character of the socially accepted principles of fairness.

Finally, Ms. Vári analysed the Finnish case in terms of a set of success criteria, derived from previous research studies on radioactive waste management. By finding that the Finnish process met the majority of these criteria, she concluded that there are a number of important siting elements that are of cross-cultural character.

Tom Isaacs, from Lawrence Livermore National Laboratories (LLNL), investigated the Finnish case from the perspective of strategic decision making. He emphasised that in addition to a well-organised programme of public participation, the following elements seem to have contributed to building public confidence: (perceived) competence of the implementors and regulators; (perceived) good intentions on the part of key decision-makers; and their willingness to change programme components to meet public demands.

Mr. Isaacs demonstrated the stepwise nature of the disposal facility development and showed that this development was not fully linear (e.g., an earlier municipal decision had rejected the concept of disposal). He also pointed out that although there has not been any provision for compensation for the host community, implementors worked out a win/win arrangement with the latter. Finally, he concluded that there appear to be certain elements that might be common to successful siting processes.

8.   Feedback from workshop participants

The FSC workshop has proven useful to all participants.

It was good for the Finnish participants in that it offered the first ever experience where all stakeholders where discussing with one another under the same roof and it allowed an overall, joint, look at the DiP process that could help identify, at least for some, what could be corrected and/or improved for the future.

The FSC participants found that the planned series of workshops in specific decision making context is indeed a fruitful idea. The present arrangements whereby all participants are made to interact with one another and express themselves are constructive and need to be retained in the future.

---

2. This concept of fairness is associated with the outcome of decision ("outcome fairness") and should be distinguished from the fairness of the decision-making process ("process fairness").

The definition, ahead of time, of issues to be discussed is also very helpful. The Eurajoki visit was found to be very necessary in order to understand the community position, as well as to grasp its natural environment. Appreciation was voiced for the presence of Parliamentarians and of the opponents, which gave a further insight of the difficulties that the system had to face. A voice was heard, however, that there did not seem to be enough outsiders and that, perhaps, the FSC still got too rosy a picture.

It was reiterated that this workshop was not meant to evaluate the Finnish programme, but rather to help others to take advantage of the experience matured so far. An outline of the Nordic model of waste management was sketched, characterised by:

- the absence of military waste;

- strong involvement of the industry;

- strong local communities having veto power;

- a relatively stable and homogeneous geology for repository siting;

- a regulator on the side of the local community;

- accepted responsibility for the country's own waste; and

- decoupling disposal considerations of the accumulating waste due to earlier decisions from considerations of the future development of nuclear power in own country.

Overall the Finnish colleagues and hosts were praised for their openness and hospitality.

BACKGROUND TO THE WORKSHOP

On 18 May 2001, the Finnish Parliament ratified the Decision in Principle on the final disposal facility for spent nuclear fuel at Olkiluoto, within the municipality of Eurajoki. The Municipality Council and the government had made positive decisions earlier, at the end of 2000, and in compliance with the *Nuclear Energy Act*, Parliament's ratification was then required. The decision is valid for the spent fuel generated by the existing Finnish nuclear power plants and means that the construction of the final disposal facility is considered to be in line with the overall good of society. Earlier steps included, amongst others, the approval of the technical project by the Safety Authority. Future steps include construction of an underground rock characterisation facility, ONKALO (2003-2004), and application for separate construction and operating licences for the final disposal facility (from about 2010).

How did this political and societal decision come about? The FSC Workshop provided the opportunity to present the history leading up to the Decision in Principle (DiP), and to examine future perspectives with an emphasis on stakeholder involvement.

Finnish stakeholders included representatives of the nuclear electric utility (TVO), the company responsible for siting, constructing, and operating the facility (Posiva Oy), national, regional, and local authorities: the Radiation and Nuclear Safety Authority; the Ministry of Trade and Industry; the Ministry of Environment; the Regional Environment Centre; and municipal government (Eurajoki), researchers from the universities and the national technology research centre (VTT), Parliament, and local opposition movements (Lovisa movement, Kivetty Movement). Foreign participants included the members of the NEA group "Forum on Stakeholder Confidence" (FSC) or their representatives and nominees. The FSC is composed of nominees from NEA Member countries with responsibility, overview, and/or experience in the field of stakeholder interaction and confidence. The FSC members may or may not belong to a governmental institution. Mainly, however, they represent the viewpoint and experience of national safety authorities, implementing agencies, R&D organisations, and policy-making institutions.

The Workshop helped provide a review of the Finnish programme, by and for the FSC, and by and for the Finnish stakeholders, and will help the FSC learn from the experiences. By gathering Finnish stakeholders, those who expressed favour and opposition, as well as observer-participants from the other FSC countries, and by implementing a highly interactive format, a joint reflection on a complex reality was achieved from which wider conclusions can also be drawn concerning stakeholder involvement in the long-term management of radioactive waste.

Posiva Oy, VTT Energy, the Finnish Radiation and Nuclear Safety Authority (STUK), and the Ministry of Trade and Industry (Energy Department), who responded to an invitation by the NEA, were the local organisers and workshop co-hosts.

The day before the workshop the FSC participants had the possibility to meet with the Eurajoki City Council, whose openness and hospitality was greatly appreciated. A presentation on the community was given by members of the City Council, including: its political structure; economy;

social composition; and relation to nearby communities, also the history of its involvement in the decision making regarding the hosting of the deep repository. Question and answers were exchanged with the FSC members. The latter could also visit, in the same municipality: the Olkiluoto nuclear power plant, where they learned about the TVO company history and plans; the VLJ repository for short-lived low and intermediate waste; and the investigation area for the ONKALO facility and potential site of the deep repository.

AN INTERNATIONAL PERSPECTIVE ON THE WORKSHOP RESULTS

C. Pescatore
NEA Secretariat

The majority of workshop participants considered the Finnish facility DiP both supportable (on the level of the host community), and legitimate (on the national level). Interestingly, several Finnish stakeholders emphasised that support for, and legitimacy of, the decisions can to a large extent be attributed to some unique features of Finnish political culture. Other participants, however, primarily the four thematic rapporteurs, expressed the view that although the Finnish decision-making culture may have played an important role, a number of siting elements of broader cross-cultural significance emerged from the discussions. Overall, by combining the indications from the roundtable discussions with those of the thematic rapporteurs and of the Finnish stakeholders, as well as the results of the first workshop[3] of the FSC, in Paris in August 2000, three major sets of success criteria can be identified:

- Criteria related to nuclear energy technologies.

- Criteria related to waste management.

- Criteria related to stakeholder involvement.

1. Criteria related to nuclear energy technologies

For the general public, nuclear power and the associated radioactive wastes are amongst the hazards that are perceived as the riskiest and that generate the greatest level of concern, a finding that is replicated cross-culturally in many settings. As a result, efforts to develop nuclear energy-related programmes are replete with conflicts, delays, and inefficiencies. Within the framework of the OECD NEA several key elements which could increase social acceptance have been investigated.[4,5]

One of the key elements is the "incremental, step-wise approach" leading to the implementation of final disposal facilities. According to a recent NEA publication,[6] this approach provides opportunities for social and political review after each step and for reversing former decisions

---

3. NEA, 2000. *Proceedings of the NEA/RWMC Forum on Stakeholder Confidence.* Inauguration, First Workshop and Meeting. Radioactive Waste Management Committee, Nuclear Energy Agency of the OECD, Paris.

4. NEA, 1995. The Environmental and Ethical Basis of Geological Disposal: A Collective Opinion of the NEA Radioactive Waste Management Committee. Nuclear Energy Agency of the OECD, Paris.

5. NEA, 2001. Reversibility and Retrievability in Geologic Disposal of Radioactive Waste. Reflections at the International Level. Radioactive Waste Management Committee, Nuclear Energy Agency of the OECD, Paris.

6. *Ibid.* NEA, 2001

or modifying plans. Since there are a number of sequential decisions to be made (e.g., identifying the goals of the programme, defining institutional arrangements, defining a waste management concept, selecting a site and a method), decision makers and implementors have the opportunity to demonstrate their competence and responsible attitude over time.

The Finnish case confirmed the effectiveness of the step-wise approach as a factor of success. However, it also illustrated that decomposing the siting process to subsequent stages is not sufficient by itself; in order to raise public confidence, each step has to be participatory and adaptable as well. Constraining public involvement to certain steps of the process (for example, excluding the public from the early stages), or not being open to modifying former decisions (for example, excluding alternative methods from further investigations during the EIA process) can be counterproductive.

Another key element of success pointed out in a recent NEA document[7] is the "separation of the radioactive waste management issue from the future of nuclear power". However, in the Finnish case the role of this element is rather unclear. Although one of the rapporteurs emphasises that "the decision on the future of nuclear power is not linked directly to the current waste decisions, helping to keep them from being overly politicised",[8] according to another rapporteur:[9] "the workshop clearly underlined the limits of any experts' attempts to totally separate the debate on general nuclear issues and the specific discussion on the identification of a specific site for the disposal of spent nuclear fuel." It is a fact, however, that the facility DiP applies only to wastes that are generated from the currently operating nuclear power plants. The high level of support for the disposal facility in Parliament (159 votes for; 3 against) is also likely to be attributed to a changing attitude to nuclear energy production within the Finnish public. This seemed to be confirmed by the presentation of Ms. Anneli Nikula (TVO) on recent public opinion data,[10] as well as the remarks of a session moderator:[11] "regardless of their position on nuclear issues, they [Finnish people] are aware of what their quality of life would be if their energy supply declined, if not ceased, for any reason. [...] The moderator considers that, if Finland were located in southern Europe, they might not have reached the stage that the rest of the nuclear world currently admires." Thus, contrary to former assumptions, it seems that in some cases, chances of acceptance for the radioactive waste disposal facility might be improved by connecting it with energy production.

2. Criteria related to waste management

- The need for the waste management programme is clearly established. A consensus is established that: the status quo is unacceptable; there is an important problem to be resolved; and the planned facility is the preferred solution to the given problem.

- The goals of the waste management programme are clear. The source, type, and amount of waste to be disposed of at the facility are well defined.

---

7. *Ibid.* NEA, 2000.

8. Report by Tom Isaacs (in these proceedings).

9. Report by Frédéric Bouder (in these proceedings).

10. Presentation by Ms. Anneli Nikula (TVO) at the FSC site visit at the Olkiluoto Nuclear Power Plant, 14 November 2001.

11. Summary of Session III Roundtable Discussions by Hideki Sakuma (in these proceedings).

- Site selection and selection of an implementation approach do not occur simultaneously. The waste management concept is identified and made widely known before site selection.

- The goal of the site-selection process is to identify a licensable site with host community support. Site-selection is a voluntary process in which communities are allowed to withdraw from consideration at any time.

- The goal of the implementation approach is to identify a licensable method with host community support. Introducing measures that facilitate the retrievability of the waste improves the chances for host community confidence and support.

- A win/win arrangement is negotiated with the host community. Benefits ensue from long-term commitment to support the community, e.g., jobs, taxes, financial compensation, in-kind support, as well as from other measures intended to offset perceived negative impacts.

- The host community is involved in decision making regarding site selection, method selection, and benefits. Local governments act as decision-making bodies, and local liaison groups facilitate public information, education and consultation.

3. Criteria related to stakeholder involvement

- The legal and regulatory frameworks are well defined and clear. They are adapted to changing social and political conditions on a regular basis.

- The roles and responsibilities of the parties (e.g., regulator, implementor) are well understood. The neutrality and independence of the regulatory authority is assured. The distribution of responsibilities and authority is checked from time to time and adapted if necessary.

- A clear, open, and transparent process is used in decision making. As a result, the accountability of the decisions is ensured.

- Responsible organisations are willing to engage in a dialogue which is perceived as a fair process, whatever the outcome of the consultation should be. They are also willing to adapt programme decisions to deal directly with stakeholder concerns and considerations.

- The political leaders in both the legislative and executive branches of the government display long-term commitment to the programme. There is also a need to have commitment of the different non-governmental actors.

Ongoing debates concerning the criteria of successful siting processes indicate that there is much more to be learned in this and other fields. The second FSC workshop designed to draw lessons from the Finnish case was an important step in this direction.

INTRODUCTION TO THE WORKSHOP

WELCOME ADDRESS

C. Kessler
Deputy Director-General of NEA

Ladies and Gentlemen,

It is my pleasure to welcome you to this workshop on behalf of the OECD Nuclear Energy Agency.

In this second Forum on Stakeholder Confidence (FSC) workshop, we have a unique opportunity to hear from, and interact with, our Finnish colleagues who have just completed a landmark process: reaching a decision in principle on a radioactive waste repository site. In fact, it provides us with a chance to explore how this decision was reached with the actual players in the decision. I hope that this workshop will be valuable for both the FSC and the Finnish participants.

The Finnish "Decision in Principle" was reached in a step-wise manner in an atmosphere of significant public involvement, especially as it was confirmed by the positive vote of the Eurajoki Municipality Council and the Finnish Parliament. I would like to reflect on the importance of increasing the use of such decision-making processes to the future of nuclear energy and the use of other high technologies.

There is no longer any question that technologies which have apparent negative aspects such as nuclear energy must be pursued with openness and transparency so that all are informed and consent to the risks involved. But, not so long ago it would have been unheard of to suggest such a dialogue was necessary. The sense of the expert community was that it had to manage these types of issues *for* society rather than *with* it. Now we have come to recognise that the effectiveness of pursuing such activities as selection of a nuclear waste disposal repository site depends on society's consent. Policies that lack societal support are policies that risk failure.

The OECD has been engaged in supporting governments in their dialogue with civil society. Beginning with the 1999, and also, the 2000 OECD Ministerial Council meetings, Ministers recognised the heightened responsibility to ensure the transparency and clarity in policymaking and asked the OECD to assist governments in improving communications and consultation with civil society or civil society organisations (CSOs). They noted that the OECD was not to become a substitute for government in relation to CSOs, but recognised that many OECD committees have CSO involvement and that this could be deepened. In 2000, the OECD established its Forum 2000 to invite CSO views on Ministerial agenda, and held one again in 2001.

The Public Management Service (PUMA) of OECD has played a particularly important role in the area for the OECD. Frédéric Bouder is here today from PUMA. PUMA is looking specifically at how to strengthen the relationship between governments and their citizens, especially in terms of treating citizens more like partners to consult and learn from in decision making. PUMA is also focusing attention on the critical role governments have in promoting social and political cohesion through enhancing this relationship with citizens. Finally, PUMA is reinforcing the importance for OECD governments to provide for transparency and objectivity in information and decision making to enhance public trust. As recently quoted in an OECD publication, one of PUMA's experts noted that "people will accept the outcome of a process that they perceive as fair, even if the solution is not one they would have chosen."

As you know the NEA has also been engaged in supporting governments in their dialogue with civil society. Our work spans our technical committee areas from safety to waste to radiation protection and nuclear development and I would like to highlight a few activities that have taken place since the first Forum on Stakeholder Confidence.

In the Committee on Radiation Protection and Public Health (CRPPH) we held a workshop in Villigen, Switzerland in February on "Better Integration of Radiation Protection in Modern Society". The focus of the workshop was to evaluate why certain activities can only be resolved through stakeholder consensus, such as site cleanup of a radioactively contaminated site when the question becomes one of "What is clean enough?" Of course, the stakeholder process must result in agreement, so there is a need for flexibility on all sides. The CRPPH is also helping the International Council on Radiation Protection to develop its new general radiation protection recommendations to be more consistent and comprehensible to the general public.

Our Committee on Nuclear Regulatory Activities held a workshop in November 2000 addressing the issue of public trust. Called "Investing in Trust – Nuclear Regulators and the Public", the conference confirmed the value of active regulatory involvement with the public, though saw the necessity of drawing a line with the public on what the role of the regulator is and is not. It also concluded that regulatory agencies will require more resources to sufficiently address public information needs. And finally, the CNRA decided to establish a working group on dealing with public and regulator interaction.

Our Nuclear Development Committee is beginning a project on "Society and Nuclear Energy". This project will involve a series of workshops leading to a major conference. The focus will be to analyse the sociological and political factors influencing public perception of nuclear energy, especially the range of risks it can tolerate. It will identify decision processes currently in place in NEA member countries and review opportunities for improved communication with all interested and affected stakeholders.

Finally, in our most recent meeting with our Steering Committee we held a special policy debate on the value of NEA's work in this nuclear energy and civil society to determine whether we should expand these activities or not, and if so, through what mechanisms. The overall response from our Steering Committee was that this has been a fruitful area for NEA work. But that much of the relationship between the government and civil society must remain between them rather than being channelled through an international organisation such as the NEA. The reason was that so much of the dialogue must be nation specific due to the strong cultural element in achieving success in the dialogue.

I would like to return to today's event and not the excellent papers prepared by our Finnish colleagues to set the stage for our discussions. They give many ideas as to how confidence was built up in their decision process among the many stakeholders in such a way that all agreed to say yes to the Decision in Principle. It will be important to recognise today and tomorrow what aspects of this process may be unique to Finland and what may be more generally applicable. I think we will find there is no recipe for success.

In closing, I would like to emphasise the importance NEA places on its role in assisting governments in understanding what gives confidence to various stakeholders in a decision process and thus help governments work more effectively and efficiently. It is our intent that fora such as these increase public officials' awareness as to what each other is doing to address stakeholder interests in public decision making, especially what has worked and what has not. The bottom line we all share is that we can develop more successful policy and decision-making processes in the nuclear energy field.

SESSION I

BACKGROUND TO THE DECISION IN PRINCIPLE

Chair: C. Létourneau
Department of Natural Resources, Canada

EXPLAINING THE RATIFICATION OF NUCLEAR WASTE DISPOSAL BY THE FINNISH PARLIAMENT: POLITICAL CULTURE AND CONTEXTUAL FACTORS

I. Ruostetsaari
Department of Political Science and International Relations
University of Tampere, Finland

According to the *Nuclear Energy Act* the government's Decision in Principle (DiP) on the nuclear waste disposal needs to be ratified by Parliament. The DiP was ratified by general consent (159-3) on 18 May 2001. How we can explain this parliamentary consensus taking account that the previous DiP concerning construction of a new nuclear power plant was overruled in 1993 and the public debate on nuclear power is still pronounced. The explanation can be sought, together with the institutional arrangements, from two sources; on one hand from the Finnish political culture, i.e., traditional and inherited ways of decision-makers to make decisions and citizens' ways to react to those decisions, and on the other hand, from current contextual factors linking to nuclear waste management.

Finnish political culture has been characterised by five factors (see Nousiainen 1992): Nordic concept of autonomy (municipal self-government versus the State); Lutheran religion (allegiance to law); Capitalism; Nationalism; Power of the ruler and the bureaucracy including: State-centredness of the Finnish society; Consensual style of decision making; Working of Parliament is directed effectively by the government; Citizens' regard for governmental institutions; General trust on technology in order to settle societal problems and; Citizens' slight civic competence.

Contextual factors (hypotheses for the discussion) which effected on the process of nuclear waste disposal:

- Exceptional nature of the DiP in Finnish legislative tradition, "general interest" versus "the overall interest of society".

- According to the amendment of the *Nuclear Energy Act* (1994), all radioactive waste produced in Finland must be handled and disposed in Finland → few alternatives.

- The nuclear waste disposal was not politicised; the symbolic nature which characterise nuclear power was not linked to nuclear waste disposal.

- Grand governmental coalition (the Greens together with the Socials Democrats, the Conservatives, the Left Alliance and the Swedes Speaking Party).

- Nuclear waste disposal was not polarised between the government and the opposition.

- The resistance of the civic movements stayed on local level.

- The preparation process of the nuclear waste disposal was open and there were arenas for public debate (the Environmental Impact Assessment).

- Actors' adaptability under the critics (the government: disposal of nuclear waste produced by the potential fifth nuclear power plant will be resolved separately -> the nuclear waste management and the construction of the fifth nuclear power plant was not intertwined; Posiva: a amendment to the original retrievability plan of the repository.

- Autonomy of the candidate municipalities were appreciated by actors, there was a competition between two municipalities concerning siting the repository; Vuolijoki-agreement between Posiva and Eurajoki municipality.

- Role of media debate was slight: the nuclear waste management was primarily seen as a technical safety issue.

HISTORICAL AND LEGISLATIVE FRAMEWORK OF THE DECISION IN PRINCIPLE

J. Manninen
Ministry of Trade and Industry (MIT), Finland

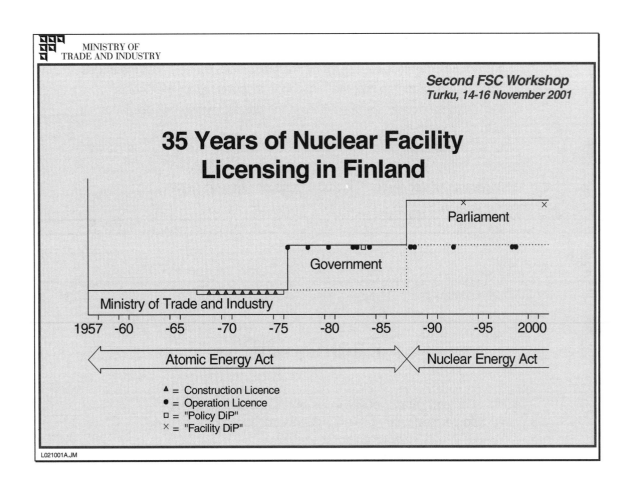

"Policy DiP"

- Decision in principle on the objectives to be observed in carrying out research, surveys and planning in the field of nuclear waste management (November 1983):
  - % … one suitable site where the final depository [for the spent fuel] can be constructed, if so decided, has been selected by the end of the year 2000

- NEnA of 1987:
  - % a licence-holder responsible for all nuclear waste management measures and there appropriate preparation, and is responsible for their costs;
  - % the MTI shall decide the principles on the basis of which this obligation is to be implemented;
  - % STUK is responsible for the supervision of the safe use of nuclear power

- * 1983: year 2000 far ahead % in 2001: decision made in 1983

"Facility DiP"

- at an early stage; political decision; general framework
- wide discretionary powers: "the overall good of the society"
- pre-EIA % a town plan not yet required,

- the veto right of the municipality
- STUK's powerful position
- involvement of the Ministry of Environment (established 1984)
- general hearing; a description of the facility, env. effects, safety

- **"double negation"**
  = "no factors indicating a lack of sufficient prerequisites"

Facility DiP %some problems

- Act on Environmental Impact Assessment (1994)
    - %   interfering with the "pre-EIA" of the DiP
    - %   different deadlines and alternatives → multiple hearings

- Nature of the DiP
    - -   is a facility specific decision
    - -   understood as a general decision on waste management

- Originally tailored for nuclear power plants

Post DiP

- Even more clearly than up to now:
    - %   not a national undertaking
    - %   different roles of the stakeholders
    - %   STUK's powerful regulatory role

- "Policy DiP" still defining the general framework
    - %   MTI refining details, i.a. the timetable

SAFETY: ONE CRUCIAL ELEMENT FOR DECISION MAKING

J. Laaksonen
Director-General Radiation and Nuclear Safety Authority (STUK), Finland

The objective of the nuclear waste management is to deal with the waste in a manner that protects human health and environment in an equal manner both to day and in the future. Nuclear waste has to be isolated from biosphere for extremely long periods of time, and it must not impose any kind of undue burden to future generations. Safe final disposal is therefore an inevitable element in waste management.

Reaching a decision on final disposal of nuclear waste requires a multidimensional approach where protection from radiation in the short and in the long term is a necessary but not the only matter to be addressed. Other safety related issues are safeguarding of nuclear materials, and physical protection of the materials and facilities. The societal dimension involves public's role and confidence in the disposal process, democracy in the decision making, funding of the disposal, and 3rd party liability. Also timing of the decision-making process is most important if one wants to ensure high level of safety and to avoid unnecessary costs.

All of the above topics are interrelated and their role and weight may vary from country to country. The climate and the geological and geographical circumstances in each country set certain conditions, but should not pose major obstacles for finding an acceptable disposal solution within each country. Other issues with importance are the available safety technical competence and financial resources, as well as the size of the nuclear programme. In any case, a key for making the necessary decisions at the right time seems to be the national culture. Important features of the culture are the emphasis on safety aspects in general, and the ability to reach a political consensus on difficult matters. However, in any national approach a high priority on safety is a necessary condition to succeed in the disposal of the waste.

The Finnish Nuclear Energy Law sets some main boundary conditions for dealing with the nuclear waste. In practice, these conditions imply that all nuclear waste we have produced must be handled within the country. No foreign waste can be imported. The closed fuel cycle options are in practice excluded, and the only feasible option for spent fuel is disposal as waste.

General safety considerations have led to a conclusion that in our circumstances and with current technology the best alternative is the disposal of spent fuel deep into the bedrock. All safety goals and technical safety requirements have been issued with this approach in mind.

For safety, regulations have been issued rather lately. The regulations consist of the Decision No. 478/1999 by the government, and STUK's YVL-guides 8.4 and 8.5 (draft). The issues covered include: radiation protection objectives and principles for short and long term, also including anticipated abnormal events and postulated accidents; disposal facility and its operation (safety principles, elimination of various hazards, safety evaluation); long-term safety (redundant barriers, geological site characteristics, depth); specific requirements for underground facility (design, integrity of host rock, prevention of damage to waste canisters); and demonstration of compliance with safety requirements (operational safety, long term safety, handling of uncertainties).

Means for safe disposal of spent nuclear fuel have been planned and studied based on a government decision in 1983. This decision gave the general objectives and the schedule for the entire process of spent fuel management, from the interim storage until the final disposal.

In the course of about 15 years, a large number of plans and studies were provided by the nuclear industry. Several updates of overall safety analysis were made, as additional supporting knowledge had been accumulated. STUK reviewed these analyses and gave its judgements respectively.

When the field studies had been completed on several candidate sites to an extent that was feasible from the surface and from small diameter bore holes, it was time to apply for a decision in principle (DiP) from the government. This is the first step in licensing of the final disposal facility, and it is required before major investments can be started. One of the necessary conditions for making the DiP was a positive statement by STUK, based on its preliminary safety appraisal of the disposal concept.

In the DiP stage, no definite conclusion on safety of the proposed disposal concept is required. Thus STUK was only expected to state that nothing had been found which would raise doubts about the possibility to achieve the required safety level.

The government of Finland issued the DiP on 21 December 2000, and Parliament endorsed it on 18 May 2001. The DiP means that the site has been chosen, and there is a firm commitment by the respective municipality to host the disposal facility. It also means that safety studies can be continued deep underground in the planned location of the disposal facility.

The planning and research work towards safe disposal will now continue, even more intensively than before the DiP. Safety related issues, which are to be pursued, include the following:

- Technological area: fabrication, sealing and inspection of the Cu-capsule, disposal techniques and materials, transport arrangements, safeguards.

- Barriers: fuel matrix degradation, capsule corrosion, behaviour of bentonite during the early transient conditions and in the long term, backfill materials including their chemical impact to the ground water.

- Site investigations at Olkiluoto: establishing the base line condition of the bedrock, underground rock characterisation facility, R&D underground.

- Safety analytic: update of current safety analysis.

It is expected that it will take at least ten years to receive such detailed assurance of safety that the spent fuel final disposal project arrives at the next important milestone, the construction permit. Also then, safety will be one crucial element for decision making.

SESSION II

THE PROCESS OF STEPWISE DECISION MAKING IN FINLAND: PAST AND FUTURE OF THE DECISION IN PRINCIPLE

Chair: J. Lang-Lenton
ENRESA, Spain
Moderator: M. Aebersold
Bundesamt für Energie (BfE), Switzerland

THE PAST AND FUTURE OF THE DECISION IN PRINCIPLE:
THE IMPLEMENTOR'S POINT OF VIEW

V. Ryhänen
Posiva Oy, Finland

Finnish legislation for use of nuclear energy prescribes the responsibility for nuclear waste management to the waste producers. TVO, the utility operating the Olkiluoto power plant, launched already in the early 1980s research and development activities for final disposal of spent fuel. Also collection of funds for the future waste management was started by the utilities in a very early phase of power plant operation. In 1995, the owners of the Olkiluoto (TVO) and Loviisa nuclear power plants (Fortum Power and Heat) established Posiva, a jointly owned company to take care of deep disposal of spent nuclear fuel.

When the operation of the Finnish nuclear power plants was started in the late 1970s, the operation licences included requirements for safe management of nuclear waste. In 1983, the government decided on the major milestones and schedules for the national waste management programme. Site selection in 2000 and commencement of disposal operations in 2020 are examples of milestones already from that time. The decision from 1983 formed an essential basis for the planning of activities needed for the development of a final disposal facility in Finland. Long-term programme and stepwise construction of the waste management facilities (e.g., final repositories for low- and medium-level operating waste) were important factors that have also supported recent successful decision-making process for spent fuel disposal.

Now, about 20 years have been spent creating technical and scientific basis for final disposal. The planned facility for spent nuclear fuel consists of an encapsulation plant and an underground repository to be constructed in crystalline bedrock at a depth of about 500 metres. Copper canister is an essential engineered barrier for safety of final disposal. In the middle of 1980s the siting programme proceeded to the phase, at which the activities were to be focused on selection of candidate areas and to the start of deep drillings. Already then it was evident that promoting of public acceptance and good co-operation with the host municipalities would be necessities for the success of the project; according to the *Nuclear Energy Act* a precondition for site selection was an approval of the host municipality.

In the 1980s, communications were integrated as a part of the siting programme and joint liaison groups with the municipalities were formed. In the beginning of the 1990s, it was still an open question, whether local support could be created for the project. However, towards the end of the last decade, development of public opinion, especially in the host municipalities of the nuclear power plants, was favourable. Public involvement as a part of the Environmental Impact Assessment process played an important role at local level in 1997-98, when the programme was approaching the decision-making process. Assessment and comparison of alternatives in relation to different ethical, environmental and technical aspects, also including non-implementation, was very important in national decision making.

In addition to the organisations responsible for waste management, the Finnish society has been committed to development of safe solutions for nuclear waste management, including spent fuel disposal. The measures taken by the government and the regulatory bodies have been of primary importance. The laws, regulations and safety criteria have been developed and updated according to the progress of the deep repository programme. The government's early activities in setting the main targets for the long-term programme have helped the implementor to continue the necessary geological investigations and other siting efforts without too strong dependence on the changes of public opinion during the long investigation period.

For the implementor, the Decision in Principle, which was ratified by the Finnish Parliament in May 2001, means that the development of final disposal can be continued on the basis of the planned repository type, and that the future investigations can be concentrated at Olkiluoto in Eurajoki. The next step, when proceeding towards the year 2020, will be construction of an underground rock characterisation facility. This facility will be essential in confirming the properties of the selected site for construction licence of the repository. Furthermore, it will give Posiva possibility to learn about underground operations when proceeding towards the construction of the repository in the 2010s and later for operation phase.

THE REGULATOR'S RELATIONSHIP WITH THE PUBLIC AND THE MEDIA: SATISFYING EXPECTATIONS

T. Varjoranta

Director, Radiation and Nuclear Safety Authority (STUK), Finland

Final disposal of spent nuclear fuel into the Finnish bedrock has been studied more systematically in Finland since 1983. The work, covering siting, technical plans and safety research and analytics, has been based on the decision made by the Finnish government. The responsibility, both operational and financial, of the waste management, including disposal, lie with the waste producers who have formed a joint private waste management company Posiva Ltd. Radiation and Nuclear Safety Authority, STUK is the independent safety regulator.

Already a few years ago STUK became concerned about results of some studies indicating that the knowledge of the general public about safety related issues of the final disposal was poor and frequently incorrect. Taking also into account that the host municipal has legal veto right in the siting process of the final repository for spent fuel, STUK considered it important that especially the local people and decision makers have correct information, understand that information and their attitudes would willingly emphasise safety. Therefore, STUK initiated a profound study in the Helsinki University with the following objectives: to find out if the local people, decision makers and media wish STUK were to play a role in communication, and if so, what they expected from STUK. The main results of the study showed that a proactive "referee" role by STUK was appreciated. The results also revealed how differently from experts the local public viewed risks, how local people and media were interested in very pragmatic everyday safety related issues rather than the long-term safety challenges that kept experts occupied.

A programme for co-operation and direct communication with the public media, including oral and written materials, seminars and discussion meetings, was established and carried out by STUK. The programme and activities of STUK were solely based on the studied needs of the local public and their representatives (elected ones, municipal administration, civic and environmental organisations) which they communicated to STUK. STUK representatives from the Director General to inspectors were a frequent guest in communities, as well as in local and national media. STUK's main objective in this area is to build credibility and public confidence in the high quality and transparency of the decision process of the disposal project itself. The objective is not to gain public acceptance as such for disposing the waste.

The co-operation programme has been ongoing now for some years. The reception of STUK in municipalities where site investigations have been carried out and in the local media has been positive. Safety issues have been discussed and handled in a passionless and businesslike manner. Issues have been relevant. STUK intends to continue the proactive interaction with the local public and media based on their needs and transparency.

STUK has positive experience about its information process to the Finnish public. However, we do not believe that what has worked well in Finland would necessarily work well in another country. We do not believe that there is a general way or a global method to carry out a successful public information process; this challenge has to be approached on a basis which recognises local needs, dialogue and media cultures specific to the area and country in question.

For STUK the key success factors have been:

- Main resources are focused on municipality level (less on national or international level), including visible participation of STUK's highest management.

- All activities are based on the needs of the municipality.

- STUK acts proactively on municipality's side and promotes confidence in the disposal process.

- STUK emphasises the importance of the domestic safety-technical competence in the final disposal process and that safety can be best assured if the disposal R&D process continues without interruptions.

- Attitude of trust and help to the media to support them to make quality, many-sided reviews and articles about final disposal issues.

- Transparency towards public as well as scientific community in all issues and details of the final disposal project.

- Providing promptly best available information to the public and media.

THE OPPOSITION'S TURN TO SPEAK

K. Tuikka

Representative of "Kivetty liike" (opponent group to the investigation site), Finland

I am Kimmo Tuikka and I work as a teacher in Äänekoski. I have lived there since 1969 except my studying time 1985-1991. As I am interested in environmental issues, I have participated in the action of the local association for nature conservation, *Ala-Keiteleen luonnonystävät*, since 1992. Working against nuclear waste disposal into the Kivetty area had been one part of the activities of the association since the site studies commenced at the end of the 1980s.

The situation clearly changed when Kivetty was chosen to be one of the three places where the studies continued. The municipality of Konginkangas, where Kivetty was situated, was united with the town of Äänekoski at the beginning of 1993. Since then the anti-nuclear waste action was separated from the action of nature protection association and the Kivetty-movement was enlarged to the neighbouring towns and municipalities.

The main working themes of the Kivetty-movement were to make the opposite views for the TVO's (later Posiva's) information known and to alert people of the issue. We questioned the safety of transport and storage of the radioactive waste. We were also worried about the image of large areas around the planned disposal site and also the social effects. In our work we co-operated with other environmental movements in Finland: The Finnish Association for Nature Conservation; Greenpeace; Friends of the Earth; and, the Romuvaara-movement in Kuhmo, one other studying site.

Many of the authorities and head politicians of Äänekoski were either, for the nuclear disposal site or otherwise, they stayed quiet. Fairly soon we noticed that we had to go in for communal decision making if we wanted to end this silence. So we gathered a group of ten anti-nuclear-waste persons and took part in the communal election in autumn 1996. In the election we co-operated with the Green league of Finland, which had previously not had representatives in the town council of Äänekoski. During the election four of us passed through and historically changed the power-relationship within the town council from the former socialist majority to non-socialistic.

When the Environmental Impact Assessment (EIA) started, we were a bit suspicious about its independence from Posiva, because Posiva (and earlier TVO) had used lots of money on advertising its work and had left us with a feeling that they would try to buy acceptance from the public. Why would they not try to buy the "right" answers to their questions also?

As the EIA studies went on, I was interviewed sometimes as an activist and as a member of the town administration. Discussions with the interviewers left me with the feeling that our opinions were valuable and that the views of the local nuclear waste supporters were not always realistic for the secondary economical impacts of the plan. The EIA process also handled other issues apart from just economical and technical, which had been on the focus earlier, for example in the information leaflets of Posiva.

I was very surprised to notice that the results of EIA supported our opinions on the suitability of Äänekoski for nuclear waste storage: minimising of the transportation risks; effects on the communal economy; and a principle, that it is not only techniques and economy, but also larger social effects. The plan would also be so large, that it would need a larger infrastructure than a distant place in the periphery could afford.

There are still some questions, that I would like to be answered: Was the decision to choose Olkiluoto as the only investigation site done short-sightedly in order to make it possible to go on with plans of a new nuclear plant in Finland? Is the planned technique and depth of the disposal safe enough for the environment and mankind? I also have a little suspicion that some day nuclear waste will become a commodity in the European Community and will be disposed in the remote, sparsely populated areas, perhaps in Finland.

THE PROCESS OF STEPWISE DECISION MAKING IN FINLAND: PAST AND FUTURE OF THE DECISION IN PRINCIPLE

A. Lucander
Member of Eurajoki Municipality Council, Finland

The municipality of Eurajoki is located on the west coast of Finland 15 km from Rauma to the north. The two Olkiluoto Nuclear Power Plant Units are located on the isle of Olkiluoto from where the distances to the centre of both Eurajoki and Rauma are roughly the same. The population of Eurajoki is today a little less than 6 000.

The role of the municipality is very important according to the Finnish law. According to the *Nuclear Energy Act*, the acceptance of the site municipality for a nuclear project is a precondition before the national government can accept it and before Parliament can ratify it.

In the early 1970s, private industry was looking for a proper location for a nuclear power plant in Finland. The Isle of Olkiluoto with a size of about 900 hectares was chosen as the best alternative. The State of Finland owned 80% of the area.

At Eurajoki, the "nuclear era" started in December 1973, when the municipal council of Eurajoki unanimously agreed upon the construction plan of Olkiluoto. One of the council members expressed a dissenting opinion by stating that "I do not accept the nuclear matter at all." As a precondition for the acceptance was a requirement that spent fuel shall be sent abroad for reprocessing according to the international practice at that time.

After that, an intensive period of construction followed. The people of Eurajoki got jobs in ample and the income were good. The municipality got taxes even from the companies building the plants. In 1978, an operating licence was applied for from The Ministry of Trade and Industry. At the same time, the municipal council was informed about it. Also the matter of spent fuel was discussed. The intention was to send the spent fuel abroad for reprocessing, while the residual high-active waste was to come back for storage.

In 1983, a target schedule was set forth by the government. According to it TVO has to make preparations for the final reposition of the spent fuel from the Olkiluoto nuclear power plant deep into the Finnish bedrock. From the point of view of people living at Eurajoki, final disposal solution has thus been under preparation for almost 20 years. About three years were used for screening and identification of potential investigation sites. Preliminary investigations were started in 1986 at five sites including Olkiluoto. More detailed investigations started in 1992.

Before site investigations started at Eurajoki, the municipal council had taken a strict negative position against final disposal of spent fuel in the municipal strategic plan. It stated that the municipality of Eurajoki will act so that high-active spent fuel shall not be deposited at Eurajoki. A very tight debate and voting about this clause did take place 1993. The result was 13-13 (1 abstained from voting). The clause remained with the chairman's vote. Next year municipality council discussed this topic for two hours and had 30 addresses. Finally, it was deleted with votes 15-10 (2 abstained from voting).

In Finland, taking care of nuclear waste has been a responsible activity. It has been from the preparations in early 1970s and it has to be responsible co-operation, where partners are the legislator, local decision-makers and those, who implement it in practice. Important milestones in nuclear waste management, which the local decision-makers had to take into consideration, were among others:

- 1983: the government's decision about time schedule for final disposal of spent fuel.

- 1994: the Nuclear Act was amended by Parliament so that since 1996 nuclear waste could not be exported from Finland. At the same time, importation of nuclear waste to Finland was prohibited.

- Environment Impact Assessment (EIA) legislation.

- Founding of Posiva Oy.

Preparations for forming the opinion of the municipality on possible application for a decision of principle started roughly five years ago. Partners in this work have been various groups, such as: co-operation group between the municipality and Posiva Oy; co-operation group between the municipality and TVO; and preparing group for comments. The environment impact evaluation process itself has been a versatile producer of information, where opinions of the local population and decision-makers were charted several times and from all sides.

Numerous investigations were undertaken. Above all economical impacts of the project in the municipality were investigated, as well as the health and social impacts. The environmental impact process created a discussion platform in which all interested inhabitants could participate, express their opinion about the project and receive answers to their questions.

Results of research and investigations have been presented in open EIA-seminars.

All statements concerning the application for the Decision in Principle were presented to the decision makers of Eurajoki. The statements were in general in favour of the project. Some of the neighbour communities criticised the project.

Final acceptance from Eurajoki was given in the municipal council meeting on 24 January 2000. Much publicity was given by the press and TV to this occasion, where the council members had one hour's pertinent discussion, where opinions of the counterpart were respected. The result was 20-7 in favour of the project. Those with a negative attitude feared possible risks for the environment and too big concentration of nuclear activities. They also had fears that a spent fuel repository would have a negative influence on the image and that foreign spent fuel might be imported to Finland. Those with a positive stand emphasised the effects on employment and economy and further that Eurajoki having the benefits from nuclear power production has also the moral responsibility of wastes and that the safety authority STUK is a competent and responsible supervisor of the safety. They were even of the opinion, that the project has a positive influence on the image. Favourable opinion by the municipality was a necessary condition for the government's and Parliament's Decision in Principle. At this stage, it gives permission just to start building of a rock laboratory for final investigations. Later, construction and operating licences, on which the municipality does not have any veto, will be needed. We at Eurajoki have taken for granted that only spent fuel produced in Finland may be deposited at Olkiluoto and that we shall be informed in future about the results of all investigations.

SESSION II REPORT
THE PROCESS OF DECISION MAKING IN FINLAND:
ITS HISTORY AND OUTCOME

Moderator: M. Aebersold
Bundesamt fur Energie (BfE), Switzerland

The roundtable discussions focussed on three questions.

1. **What were the most important steps in the decision-making process for the different stakeholders?**

   - A key element of the Finnish programme was the 1983 governmental decision setting the major milestones and the schedule to measure the achievement of these milestones. This defined at a very early stage the institutional structures.

   - In 1994 a governmental decision prohibited the export of waste for disposal outside the country. This was a major milestone and a clear sign to proceed with a "domestic" disposal solution in Finland.

   - The Environmental Impact Assessment (EIA, 1997-99) permitted the open discussion of important political themes on waste management such as direct disposal against reprocessing, and the export of SNF.

   - STUK's Safety Review and the government's Decision in Principle (DiP) of 2000 were as important milestones as was the approval of the DiP by Parliament in 2001.

2. **What influenced the process and the outcome?**

2.1 *Institutional framework*

Most importantly was the institutional framework that provided a timetable for the waste management programme. A step by step decision-making process was put in place at an early stage. This was facilitated by a simple organisational structure that clearly assigned responsibilities (i.e., regulator, implementor). In addition, the political decision to prohibit the export of SNF and the early introduction of "geological disposal" to manage waste, played an important role.

2.2 *Acceptance*

There was a broad political consensus, on a national as well as regional level. Moreover the local municipality was involved in the decision-making process at an early stage. As the municipality was already a nuclear site, the residents were familiar with nuclear matters and had a more sophisticated technical know-how. It also helped that the regulator and the implementor were known to the local government and the public and therefore the building up of confidence did not start at zero. It was also agreed that the fact that the different sites were in competition had a good influence on the acceptance.

2.3   *Participation and transparency*

A well-defined decision-making process, such as the EIA, allowed for the participation of interested stakeholders. The implementor and the regulator recognised the importance of an open and transparent process and took the necessary measures.

3.   What are the lessons learnt?

It is widely recognised by a majority of the stakeholders that a problem exists and that the problem needs to be solved and it can be solved. This can only be achieved in an atmosphere of confidence and trust. STUK is widely accepted as a competent regulatory body with the necessary technical expertise. Furthermore, the implementors (TVO, Posiva) have accepted their responsibilities from the very beginning.

The municipality is a major stakeholder in the decision-making process. Holding the power to veto made it easier to say yes.

It must be understood that there are differences in perception and expertise between the "experts" and the "laymen". This can be dealt with by taking into account of public concerns through open and transparent information policies.

In conclusion, participants at the roundtable discussions all agreed that the following factors were key to the Finnish Waste Management programme:

- DiP as part of a stepwise procedure and as principal decision for implementation.
- EIA as a structure and guide for public involvement and participation.
- STUK as a regulatory body which creates confidence.

SESSION III

STAKEHOLDER INVOLVEMENT,
PARTICULARLY IN THE ENVIRONMENTAL IMPACT ASSESSMENT

Chair: P. Ormai
Public Agency for Radioactive Waste Management (RHK), Hungary
Moderator: H. Sakuma
Japan Nuclear Cycle Development Institute (JNC)

STAKEHOLDER INVOLVEMENT PARTICULARLY IN THE ENVIRONMENTAL IMPACT ASSESSMENT: FINAL DISPOSAL FACILITY PROJECT

A. Väätäinen

Senior Adviser, Ministry of Trade and Industry, Finland

The Ministry of Trade and Industry is the competent authority in all nuclear facility projects. Consequently, one of the duties of the Ministry was to organise a public hearing process both on the EIA Report and, separately, also on the application for a Decision in Principle concerning the final disposal facility project. A public hearing in this context means a procedure of some months' period, when anybody has the right to present opinions on the current issue to the Ministry.

The EIA process itself was preceded by the EIA programme (in 1998), and the programme stage also included a hearing process and the Ministry's statement on it. This presentation, however, is confined to dealing with the Ministry's role and experience of the EIA stage itself and especially the hearing process on the EIA Report.

According to the *Nuclear Energy Act*, a report of environmental impacts must be annexed to the application for a decision in principle. An essential point to be noted is that no decisions on the project are made during the EIA. The EIA procedure ends when the competent authority, i.e., the Ministry of Trade and Industry, has provided its final statement on the EIA Report's adequacy.

The table below illustrates the public activity in the hearing process. The most of the public opinions sent to the Ministry during the hearing on the EIA Report were not restricted to the current process (EIA) only, but included also views on nuclear issues in general or opinions on the decision making concerning the final disposal project. This may be due to the partial simultaneity of the hearing on the EIA Report and the hearing on the application for a decision in principle. All opinions and points of view were brought to the government's knowledge irrespective of the formal context, in which they had been addressed to the Ministry.

Amount of opinions sent to the Ministry of Trade and Industry in the public hearings on the EIA Report and on the application for a Decision in Principle:

	Civic and environmental organisations	*Individuals*
EIA Report concerning four sites (Olkiluoto, Loviisa, Romuvaara, Kivetty)	4	15
Application for Decision in Principle Site: Olkiluoto	0	100

The role of the EIA in decision making:

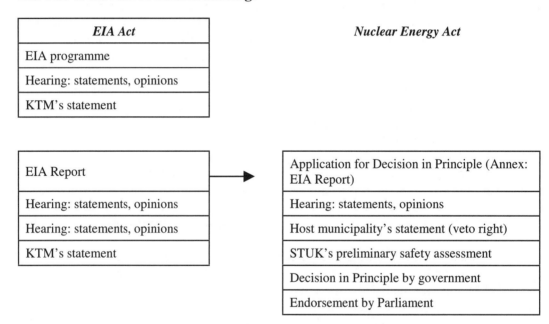

From the Ministry's point of view both the EIA process and the official public hearing process on the EIA Report can be considered successful. According to enquiries and the impressions of the Ministry people thought, that the amount of written material, contacts and information they were provided with during the EIA, was sufficient. Furthermore, on the ground of some opinions sent to the Ministry it can be supposed, that it was demanding and partly problematic for the public to adopt and manage the abundant information coming from different sources or to distinguish reliable information and research results from unreliable information and groundless opinions.

The *EIA Act* came into force as late as in 1994 and was now for the first time applied to a project where a completely new nuclear facility is planned to be constructed. Consequently, the question of the role of the EIA and the relation of the EIA legislation and the nuclear energy legislation as well as the question about people's possibilities of influencing the project plans by participating in the EIA process came up.

PUBLIC PARTICIPATION IN THE ENVIRONMENTAL IMPACT ASSESSMENT: ONE ALTERNATIVE OF INVOLVEMENT

P. Hokkanen

Department of Political Science and International Relations, University of Tampere, Finland

Posiva's EIA for the final disposal of nuclear waste covered four candidate municipalities, Eurajoki, Kuhmo, Loviisa and Äänekoski, where the possibilities of final disposal of spent fuel were being investigated. The implementation of the EIA was a comprehensive process in many ways, when considering the history of EIA in Finland. There was an "EIA era" for almost three years in all candidate municipalities. The EIA process was seen in the everyday life of the municipalities. The EIA process has been dubbed of "the EIA of the century" in Finland. The central political aim of the EIA – to increase participation – moreover brings the question of nuclear waste into a new arena.

The Environmental Impact Assessment Act (267/1999, 1 §) underlines public participation. There are many ways of public participation available at the local level. Some are "direct" and some "representative" in nature. For example, in the case of the final disposal of nuclear waste, local inhabitants have had a number of opportunities to take part in and to influence the ongoing process. In the EIA process of the final disposal there were three ways to participate: Public hearings (and other meetings) before and after the EIA programme and report; written opinions to the competent authority (KTM) after the EIA programme and report; and direct contacts to the EIA contact persons of the candidate municipalities.

Public participation was slight when evaluated in a quantitative fashion. Moreover, participation levels decreased throughout the process. In the municipalities, the designated EIA contact persons remained under utilised. Furthermore, the levels of public participation that did occur were themselves predicated on a small number of multiple contributors. This group of people formed what essentially became an elite group of direct participants. At least six reasons can be condensed for the low levels of participation:

- The tradition of the institutional public participation has historically been based on representative democracy in Finland.

- As an instrument of participation in the EIA is thus a novel form of political engagement.

- In general, the EIA process is felt to be ineffective. The relationship between EIA and the decision-making process is usually unclear, at least for the general public. Some people may feel that it is useless to take part in EIA, believing that there are other ways of participation outside of EIA.

- The final disposal of nuclear waste is an exceptionally long process. Tiredness and exhaustion is clearly one reason for the weakness of public activity.

- The long duration of the process itself, several public meetings and massive information activities increased the knowledge of local inhabitants about nuclear waste management and the final disposal of nuclear waste. In such a situation it is not necessary to take part in the EIA process, at least not many times.

- The amount of participation is also explained by resources for participation and their uneven accumulation. The typical participant in EIA is selected from the large mass of citizens with the result that the basis of participation narrows, and groups of few activists result.

What was perhaps even more concerning was the fact that inhabitants and policy makers from each of the candidate municipalities did not met during the EIA process. There were so many arenas of participation potentially available that the EIA was not always the most effective forum in many cases. Activity outside of the formal EIA process was thus important for all actors. This was especially so for those who opposed the plan, as the logistics of the EIA process was complicated to say the least. On the one hand it could be argued that, it was important to use such an instrument, though on the other, it is obvious that a lack of both understanding of, and confidence in the process existed.

At all events, the EIA did provide a useful mechanism for controlling and alleviating the worries, hopes and arguments of the citizens, and an efficient route for disclosing them to the authorities and the developer. According the *EIA Act*, the main aim of the EIA process is to draw inhabitants and other participants into the participation and encourage them to voice queries. For that reason, the EIA of final disposal was one of the most important instrument as well the phase of process for public participation when examined nuclear waste management in Finland in its entirety.

STAKEHOLDER INVOLVEMENT, PARTICULARLY IN THE ENVIRONMENTAL IMPACT ASSESSMENT

J. Jantunen

Senior Adviser, Uusimaa Regional Environment Centre, Finland

Summary

The statements of the Finnish environmental administration show that the EIA itself was an important stage for discussion on the selection of a final disposal facility for spent nuclear fuel. Nuclear safety and human impacts were considered to be the most important issues. The EIA process was a tool that could be used to carry on organised discussions. The Finnish environment administration was satisfied with its participation in the process, it got the information it needed and its viewpoints were seriously taken into consideration.

During Posiva's environmental impact assessment, I was working for both the Uusimaa Regional Environment Centre and for the Ministry of the Environment. Loviisa, one of the final disposal sites for spent nuclear fuel, is situated in the Uusimaa region. To present an overall view on the process in this workshop, I reviewed all the EIA statements given by the State environmental administration. The viewpoint of the different levels of the environmental administration varied.

To many environmentalists, all nuclear projects are very sensitive and there is much critical thinking about nuclear energy. However, where the EIA was concerned the attitude was quite cool. We tried to concentrate on the EIA procedure and on the requirements in the EIA legislation, and in other environmental legislation concerning impact assessment. Before starting the EIA, Posiva was in contact with all branches of the environmental administration.

The Regional Environment Centres had a very practical viewpoint. In their statements they addressed the local environmental aspects: nature, landscape, urban structure, land use planning, bedrock, groundwater and so on. Most of the statements were very short, and the regional environment centres were very satisfied with the information they had received and with the EIA in general (e.g., *The EIA is comprehensive...; The studies of different alternatives have been carried out with care and objectivity.*) When Posiva informed the environmental administration of its site decision before the EIA was completed, only one regional environment centre gave a statement on the final EIA report, and the others were no longer interested in the issue.

The Finnish Environment Institute (FEI) had a more principled and critical viewpoint. In its statements, the FEI demanded more alternatives. It also pointed out that the scope of the EIA was local, but that the problem in decision making is to determine the general interest of the society and, if necessary, the participation should be wider than local. The FEI was the only one in the environmental administration to say that the EIA was goal-directed, but it did not give any explanations for that comment.

The Ministry of the Environment concentrated on the EIA and whether the procedure followed the Finnish EIA legislation. The Ministry was satisfied with the wide participation. But it was not satisfied that the EIA was not completed until Posiva appealed the decision in principle. Additionally, the Ministry pointed out that the content of Posiva's advertising campaign during the EIA was not what the EIA legislation means by furthering participation.

Participation in a transboundary context

The Convention on Environmental Impact Assessment in a Transboundary Context (Espoo Convention), concluded under the auspices of the UN/ECE (Economic Commission for Europe), was applied to this project.

The Ministry of the Environment notified Estonia, Russia and Sweden about the project, and arranged two unofficial meetings to inform these countries' representatives of the EIA process. Posiva also arranged one informational meeting in Estonia with its Ministry of the Environment.

All three countries informed Finland that they were interested in participating in the EIA of this project, although they could not see any significant impacts in their territories. All three also submitted statements, and especially in Sweden the participation was active.

STAKEHOLDER INVOLVEMENT IN POSIVA'S
ENVIRONMENTAL IMPACT ASSESSMENT

J. Vira
Posiva Oy, Finland

The application for the Decision in Principle must include the report from a completed environmental impact assessment (EIA). In general, the purpose of the EIA legislation is to bring more transparency and interaction among potential stakeholders into the planning of projects that may have a significant impact on their physical or social environment. In addition, the EIA should look at different alternatives of the project implementation and also consider the impact of not implementing the project at all, i.e., the so-called zero-alternative. For Posiva the EIA was an important consultation process for determining the basis for the continuation of the disposal project. Because of the strong vetoes that the Finnish legislation gives to several stakeholders Posiva's EIA work was focused on subjects that the public found of greatest concern. In this respect important stakeholder groups included the local people of the investigation communities and their representatives in the municipality councils, the regulators, the public administrators dealing with environmental and energy policy issues, the scientific community involved in related research as well as the whole population and their political representatives at the national level.

In the scoping stage Posiva's EIA process paid considerable attention to the interaction with the public in the communities that were candidate sites for the facility on the basis of the site selection process already started in 1983 after the government resolution. Information in the form of different reports, brochures, leaflets, videos and exhibitions was distributed to households. Liaison groups were formed in all investigation municipalities. A large number of various meetings were held to collect guidance for the assessment from different stakeholder groups both at national and local level.

In the actual assessment phase interim results from the assessment work were presented in Posiva's information bulletins and leaflets and also at the meetings of the liaison groups.

As could be expected, safety, both operational and long-term, and particularly transportation safety, was among the top concerns among local people. In addition to articulated safety concerns one of the issues most frequently brought up in the public discussion was the "image": How would the project affect the image of the home municipality or the whole region? It was evident that a multitude of issues such as prices of real estate, marketability of agricultural produce from the region and attractiveness to tourism, were at stake in the image concerns, but, most likely, the image concerns can be considered as overall negative feelings about the planned facility.

Main concerns brought up in the official statements on the EIA programme from various institutions were related to the assessment of alternatives to Posiva's proposed spent fuel disposal concept. A special topic that was introduced to discussion during the EIA process was the question of reversibility. In March 1999 the government made a decision on general safety requirements for the final disposal of spent fuel which now included the requirement that the spent fuel must be retrievable even after the repository has been closed.

Research into the long-term safety of final disposal has been carried out in Finland since the early 1980s. This work has already produced several comprehensive safety assessments since 1985 and new assessments were prepared for the particular purpose of the EIA and the Decision in Principle. However, for the purposes of the EIA the scope of the research programme was substantially extended to address other assessment topics and issues identified in the assessment programme.

A great deal of the new studies was devoted to the expected social impact. In the general approach it was admitted on the one hand that the local people were experts in matters related to their social environment and in their own perceptions of changes in their living environment, but on the other hand it was also considered that the eventual social impact of a project spanning over several decades could not be judged on that basis only. Therefore, it was decided that rather than making firm statements of the likely impact, a spectrum of possible impacts would be described and analysed. In this way the assessment sought to combine the subjective experience of local citizens with the more general information obtained from other projects of similar nature. Of course, the difficulty of this approach was that no experience of the social effects of high-level nuclear waste repositories is available and the relevance of any suggested analogy can easily be contested.

The areas addressed in the social impact assessment included:

- Effects brought about by changes in the physical environmental (related to every-day living conditions and general living standards, amenities, social well-being).

- Effects on community and population structures and infrastructures.

- Effects on local and regional economics and well-being.

- Psychosocial effects (worries, anxiety, psychic effects).

The socio-economical assessment was based on traditional input-output analyses taking into account the direct and indirect effects on local and regional economic activity. The sociological and psychological aspects of the planned activity were on the one hand discussed in terms of the outcome of interviews and opinion surveys, on the other hand based on recent scientific studies and literature. A considerable amount of effort was spent on attempts to elaborate the meaning of image and constituents thereof as it was clearly affecting people's opinion both in the likely economic impact and in social well-being.

The final assessment report was, in general, well-taken by both the public and the authorities. The criticism expressed was rather related to practical aspects of the EIA legislation in general than Posiva's assessment as such.

WHAT COULD HAVE BEEN DONE?
REFLECTIONS ON THE RADIOACTIVE WASTE BATTLE AS SEEN FROM BELOW

T. Rosenberg
Sociologist and Chairman of the former Citizen Movement Against Nuclear Waste Disposal,
Lovisa, Finland

How best to describe the three-year long battle between Posiva, the company promoting final disposal of spent nuclear fuel, on one hand, and the local resistance, mobilised in a citizen movement, on the other? A battle undoubtedly reminding that of Goliath *versus* David, i.e., rather uneven what comes to resources, influence and knowledge.

Probably one word is enough, and that is "theatre". This because so much in the whole process, especially concerning the environmental impact assessment (EIA), reminded a dramatic spectacle with everything written in advance: the parts, the complicated and stepwise choreography – and, above all, the whole narrative, from the very beginning to the (from the resisters' point of view) bitter end.

What was, then, to be done, as the outcome of it all seemed absolutely clear from the beginning? Why bother, at all? If we, as citizens in Lovisa, were doomed already, as nuclear sites usually are, nothing would, of course, change the main script. Besides marginally, i.e., by choosing the location for the disposal plant, between one of the two nuclear sites in Finland. Quite early as it had become obvious that only these two sites (i.e., Hästholmen in Lovisa and Olkiluoto in Eurajoki) were among the potential ones, despite the official game with four candidates, of which two were non-nuclear sites (Kivetty in Äänekoski and Romuvaara in Kuhmo).

This was, in short, the situation confronting all those on the local level who decided, still, to resist the location of the final disposal facility for radioactive waste in their own municipality, in this case the small town of Lovisa. A town whose citizens had, after all, always been explicitly promised not be left with the spent nuclear fuel. (Note that I wrote Lovisa with only one "i", thus pointing out one of the main characteristics of the Lovisa case, i.e., the language dimension – something I will return to later.)

In my speech, I will try to describe the spectacle as seen from below, especially from the activists' point of view. I will do this by accounting for the strategy adopted by the Lovisa movement, the citizen movement rapidly mobilised against Posiva's deposit plans, as soon as they became public the 4 January 1997.

Of course one could claim that the narrative offered above implies that the anti-nuclear waste movement was rather preconceived in its attitudes. In a way this was true. But so were, I am positive, also the attitudes on the other side of the table. This makes the theatrical metaphor even more accurate, as both parts played their roles according to a prescribed scheme. Something which made the desired confidence and trust from the very start a rather limited resource.

Much can be said about the matter of trust, of course, but let me here summarise two of the reasons why there was such a great lack of confidence and credibility from the start.

- The first reason lies in the David *versus* Goliath constellation, i.e., in the feeling that the stakeholders in this case were completely uneven, representing fundamentally different weight categories. You do not have to suspect any sort of conspiracy in order to judge the battle lost beforehand – so big and strong were the vested interests.

- The second reason lies in the division of power and the relationship between the applicant and the regulating authorities. It is for most people unacceptable especially that the applicant itself (according to the legislation) is obliged to master the EIA, but also that the regulatory mandate is given to the two state authorities so deeply involved in the nuclear industry as The Ministry of Trade and Industry and the Nuclear safety authority, STUK. This arrangement eliminated most of the confidence from the very start, at least among those familiar with the nuclear network in Finland.

As this session deals with the stakeholder involvement, particularly in the EIA-process, I will concentrate on this aspect, though it in many ways blurs the perspective. In Lovisa, drawn into the location battle on a very late stage as we were, the whole and intensive three-year campaign against the disposal was almost parallel with the EIA-process. This forced the movement to react in consequence with this, even if we considered the whole thing rather useless and frustrating, I must admit. (Observe that we were targeted also by another, almost parallel EIA, concerning a new power plant. According to the official rhetoric totally independent of the EIA on radioactive waste, but of course in many ways connected with this.)

From the beginning, it was very clear for us in the Lovisa-movement that we had to the adopt a simultaneous double-strategy, on one hand totally refusing to discuss on the premises given by Posiva as well as by the authorities (as defined both by the legislation concerning radioactive waste and the stepwise EIA-process), on the other participating in the process, in order not to be totally marginalised. (But now, looking at the process in retrospective, I am not sure whether the latter was necessary. I think we would have done quite well also by simply boycotting the EIA – perhaps even better.)

This is not the place to discuss the disposal concept offered by Posiva, not to talk about nuclear power in general. But of course the often very strong sentiments against nuclear power heavily influenced those active in the movement. If there was a lack of confidence, the reasons for this were, therefore, to be found much earlier, and were much more basic in nature.

The Lovisa movement was, however, mobilised explicitly in order to resist the disposal plant, nothing else. The board included, e.g., even members who were pro nuclear power. This deliberate narrowing of the focus did not, of course, prevent almost every public event to get stuck into a discussion about "the ancient Greeks", i.e., disputing over and over again about all the basic elements, which is, as we all know, quite frustrating for both sides. But people simply are not that rational as it may look from the EIA-technocrat's desk. Or, to put it another way: the EIA, with its over-rational logic simply does not fit here. (i.e., in contrast with Pekka Hokkanen, I do not find EIA a good thing!)

In short, Posiva's manuscript, in accordance with the main interpretation of the legislation concerning spent nuclear fuel, offered no alternative to final deposit in bedrock. The EIA, on the other hand, implied participating in a long, frustrating, and co-optative process, however scientifically camouflaged as it was – a process which from our point of view only legitimated a discourse we refused to accept. And we were all the time quite aware of this Trojan nature of the EIA; the more you got involved, the more you had to accept the agenda-setting given.

Our strategy was, as I mentioned above, double-sided in character, and based on three corner stones: broad representativeness; professionalism, especially in the media; and refusing to discuss on the premises given, offering alternatives instead.

1.   As the planned disposal of radioactive waste affected not only Lovisa but also its surroundings the movement included representatives not only from this town (with about 8 000 inhabitants), but also from the four neighbouring municipalities traditionally regarded as the Lovisa region (in sum about 20 000 inhabitants – but the movement actually included also more distant municipalities). Also in this respect the movement thereby rejected the definition made by Posiva and the legislation. From the very beginning the demand for a public referendum was raised, and then explicitly concerning not only Lovisa, but also at least the neighbouring two municipalities, too. (Lovisa itself is geographically quite small.) This demand was never put into practice, but obviously affected the discussion, well known as it was that the anti-waste disposal feelings were more pronounced in the surrounding municipalities.

A kind of referendum was, however, put into practice, as the movement as one of its first steps in April 1997 decided to publish a petition against the disposal plant, thereby giving people living in the Lovisa region a possibility to declare their standpoint with a clear and simple NO, as an alternative to the written expressions of meaning in the EIA. The collection of names continued during the whole process, and finally (in February 2000) included almost 3 900 names.

In this connection something must be said also about the language dimension, as the cultural and linguistic aspects were of significant importance, in many ways dividing as they were the attitudes concerning the question at stake. Lovisa, and the whole region, had formerly been dominantly Swedish in character, something rapidly changed by the nuclear era, beginning in the early 1970s. From that on, the attitudes towards nuclear power in general, and radioactive waste in specific, have been strongly language- and culture-related. Or simplifying it: the nuclear age in Lovisa is strongly connected with change also on the cultural level, with its loss of former privileges, tradition and language. Or as I use to say: the nuclear age in Lovisa has split not only atoms in the power plants but also the town itself and its socio-cultural climate.

Much could, of course, be said about these cultural cleavages (well known also from other parts of the world), but let me restrict myself to one statement: I am sure that the language aspect involved affected the outcome of the location contest at least as much as most of the investigations made as a part of the EIA-process! It was, in short, for the actors involved too irritating to be forced not only to play the EIA-game on many sites at the same time, but also to do it simultaneously in two languages.

2.   The professionalism of the movement implied always keeping a very serious and knowledge-based profile, in order to gain maximal credibility (especially in contrast to the former anti-nuclear movement in the region, labelled "populistic" as it usually was). This was possible by the relatively high degree of experts from various fields in the movement, e.g., journalists, sociologists, lawyers, teachers, actors etc. And, from the political field, many of the leading politicians in the different municipalities in the region.

Special attention was paid to the publicity, especially in the local media, i.e., the two local newspapers and the local radio (both of them in both Swedish and Finnish). Though many of the journalists did not sympathise with the movement's programme (in contrast with the situation in the

1970s and 1980s), we did not have any greater problems in getting publicity, and in many ways succeeded to keep the initiative, at least in the first, and perhaps most decisive, problem-defining period. As every social scientist knows (especially today, with a paradigm so dominating as social constructivism), the question is about who defines the problem, and thereby setting the agenda.

I think the Lovisa movement, though David-like in proportion, succeeded in redefining the scene of the combat, thus tilting Goliath – at least a little bit.

3.   Refusing the discourse given meant offering alternative solutions and criticising both the deposit-model offered by Posiva and the dramaturgy prescribed by the legislation and the EIA-process.

The very first, and rather successful, seminar was titled "There are alternatives!", in April 1997. It introduced an alternative concept to the final bedrock deposit, developed in Sweden (the so called Dry Rock Deposit, or DRD), described more in detail by its designers in a following seminar one year later.

We did not have, of course, resources enough to develop this theme deeply enough, but I think we succeeded at least to keep in mind that there are – and above all, should be! – alternative solutions. Especially such ones satisfying the demands concerning a morally sustainable solution to the waste problem, i.e., stressing the waste retrievability and reversibility. (This seems, by the way, to be the most important adjustment caused by the critique from the "anti-finalists" against the idea of a final deposit. Though it, for sure, has been mainly cosmetic.)

Refusing to play the part of the perfect citizen in the EIA-process meant participating in the endless chain of seminars, hearings, surveys etc. – but only half-hearted. Parallel with this participation (including writing several expressions of opinion), we all the time found it much more important to introduce alternative solutions and perspectives. One such alternative was to put the whole EIA-process in question, e.g., by arranging a seminar explicitly focusing on this matter.

After this short and rather sketchy description of the strategy adopted by the Lovisa movement, let me conclude some of the observations made.

In all the three respects mentioned above (the broad representativeness, the professionalism and the altering of the agenda) the citizen movement can be said to have been successful – even if I, due to my position in the movement, of course could be disqualified from making such assessments! We succeeded, at least partly, if not to change the definition and the agenda at stake, at least to put it seriously in question. We also partly succeeded, as mentioned above, in keeping the initiative in the media, at least on the local level[12]). But above all, we succeeded in our main and only purpose (at least

---

12. One of our major mistakes, and this goes for all the three citizen movements, was in mainly keeping to the local level – something which suited Posiva perfectly. As a matter of fact Posiva can be said, by intention or not, to have successfully followed the ancient strategy of "*Divide et impera*". The three citizen movements dominated the local arena during the first, local phase, that's true – but who cared, as nobody knew anything of all this on the national level! The discrepancy between the massive discussion and activity on the local level and the total silence on the national one was, indeed, huge.

When the process in the end of the millennium reached its second phase, and the radioactive waste-question was lifted to the political level (by bringing the Decision in Principle to the government and Parliament), we all realised – too late – that we did not have energy, time nor resources to go through it all again. We were, after all, amateurs fully occupied with our ordinary jobs, families etc. No wonder, then, that the decision ran so smoothly through the political apparatus – the fighters lied half-dead on the local battle-grounds, too tired to take part in the last and most crucial battle.

explicitly), i.e., to stop radioactive waste deposit in Lovisa. The movement therefore decided officially to put and end to its activities in February 2000, by arranging a last public seminar, analysing the whole process.

But we did not succeed – and that, at least in my opinion, is of much greater importance – in our most crucial issue, i.e., resisting the whole idea of eliminating the problem by hiding it deep in the bedrock. Passing it over to the neighbour (in this case the Eurajoki municipality) was absolutely no victory in our mind. The Lovisa movement, together with our colleagues in Kuhmo (Romuvaara movement) and Äänekoski (Kivetty Movement), never accepted the accusations according to which we were suffering the usual, i.e., "Not In My Back Yard"(NIMBY) syndrome. For instance, the Lovisa movement stated that we even prefer a prolonged temporary deposit on the ground in Lovisa, instead of a final deposit in the bedrock in Eurajoki.

But all that was, of course, in vain. The nearer the end of the EIA-process we came, the more it showed its real Potemkin character. Or, as I used to say in the first phase of the process: the mission for Posiva did not lie in finding the rock best suited for the purpose, but the community naive, brave or stupid enough (choose the word) to accept radioactive waste. Posiva itself disqualified the process by first launching an advertisement campaign in the autumn 1998, cheaply underestimating all the critical voices, and then definitely nullified the dialogue (as the researcher Matti Kojo put it in an article analysing the process) by the Vuojoki-agreement with Eurajoki in May 1999 – i.e., before the EIA-process had been brought to a close.

Or to summarise, in contrast with most of the speakers on this conference, I do not think the EIA-instrument suits this kind of a complex problem – it is far too rational in logic, expertise-driven in practice and – in this case – betraying by intention.

At last, some words about moral. The nearer the end of the play we came, the more the discussion was turned into a moral issue – but now turned up side down, at least in the opinion of all those who always had rejected nuclear power, explicitly from a moral point of view, i.e., referring to the unsolved waste problem. Now, Posiva in its rhetoric tried to convince that the moral issue at stake was to solve the problem at last, and definitely, and not to leave it to the generations to come. That is,

---

There are, still, some organisations that have been active also on the national level, such as the nature protection organisation, Greenpeace, Friends of the Earth etc., but these are not always defined as stakeholders, and thereby excluded from the scene. (This goes, by the way, also for the FSC Workshop, as the opponent side was represented only by two local activists, but no organisation on the national level – some of which I know have been openly critical against this.)

The Finnish decision concerning a disposal facility for spent nuclear fuel may look like a miracle of confidence and trust. But let us not forget that the last part of the play gives a rather biased picture of the process. From the local point of view, it looks quite different. But how to make that reality visible?

however, precisely what the anti-nuclear movement had always claimed, as the crucial and morally most weighty argument against dealing with nuclear power, altogether!

We were, therefore, trapped – and still are. A fact even more obvious today, living as we are after 11 September. But even a short glance in the annals of the anti-nuclear movement shows that we knew it from beginning, and always warned against it: nuclear power is incompatible with terrorism. And terrorism is something we never can exclude from our scenarios, at least if we are serious – whatever our leaders may say.

SCOPING AND PUBLIC PARTICIPATION IN THE ENVIRONMENTAL IMPACT ASSESSMENT: FINNISH EXPERIENCES

A. Leskinen
Diskurssi Oy, Finland

Environmental Impact Assessment (EIA) is a process that produces information for decision making. Therefore it can be described as planning. It is also a systematic way to compare alternative options.

One major lesson learned during the early years of EIA was that everything cannot and should not be assessed. The first reports of the assessments in the US were massive documents that were of very limited use in decision making. Therefore the scoping-phase was adopted. The aim of scoping is to focus on important issues at the beginning of the EIA-process.

I try to shed some light to the importance of the scoping or programming phase of the process.

1.   Scoping and participation

Internationally it is recognised, that it is the scoping phase where conflicts can be resolved if resolved they can be. At the later phases of EIA it is much more difficult for the interested parties e.g., to get their opinions about the alternatives and impacts into the assessments. Therefore it is much more difficult to handle the disputes at the end of the process.

The EIA process in Finland has two major phases. Firstly the scoping or programming phase after which an EIA-programme is presented. In the second phase the developer does its homework with its consultants and other experts and produces an assessment report, which in some countries is called the Environmental Impact Statement (EIS). In both phases public participation takes place. The public must be given opportunities at least to give written comments on both EIA-documents. In practise more effort has been put into public involvement for instance in motorway-planning.

The Finnish EIA of the final disposal facility of spent nuclear fuel was exceptional because of its participative programming phase. Most public participation efforts took place before the EIA-programme was published. The people and NGOs at the four possible sites and also nationally were given ample opportunities to get information and to express their opinions. For instance 20 special meetings with neutral facilitators were held to gather opinions about impacts that should be assessed. On the whole the approach was much more "bottom up" with people then "top down" with just experts. Still there is of course room for criticism, this and that could have been done better.

2. Scoping the options

The heart of EIA is comparing the options against the assessed impacts. At the EIA of the final disposal facility of spent nuclear fuel the alternatives were compared superficially and then scoped at an early stage. Only one technical alternative (final disposal to the bedrock) was assessed thoroughly.

Posiva carried out the impact assessments in accordance with many of the demands of the NGOs and in accordance with the statement of the competent authority. The NGOs along with other parties had an influence on the fact that Posiva did compare the alternatives and presented the reasons for scoping the alternatives. In addition, the NGOs influenced on the fact that retrievability and monitoring of the waste shall be studied in detail before granting the construction permit.

However, the NGOs opposing final disposal did not get their major demands through – the project was not postponed and other alternatives were not adopted for further studies. This might have given some NGOs reasons to blame EIA as inefficient in this case, and thus inefficient as a planning tool.

3. Conclusions

When evaluating EIA, the quality of the process is one thing and the effectiveness of EIA is another.

The EIA of the final disposal facility was of good quality in many respects. For instance the timing amount and quality of participation as well the quality of the public documents were very good compared, for example, to any of the almost 130 EIAs that have been completed in Finland.

The effectiveness of EIA is much more difficult to evaluate. I am not going to analyse effectiveness here, but it is obvious that the opinions of the parties on effectiveness differ.

SESSION III REPORT
STAKEHOLDER INVOLVEMENT, PARTICULARLY IN THE ENVIRONMENTAL IMPACT ASSESSMENT (EIA)

Moderator: H. Sakuma
Japan Nuclear Cycle Development Institute (JNC)

1.   Introduction

Owing much to the seating arrangements that allowed a range of Finnish stakeholders and FSC representatives to share eight small tables, discussions were quite active and hence produced the maximum output, given the time limitations. It should, first of all, be acknowledged that the highly interactive format chosen for the Workshop was a success.

Session III focused on the Environmental Impact Assessment (EIA) and the involvement of a variety of stakeholders therein. As introduced by a majority of the plenary speakers prior to the discussion, the EIA formed an integral part of the application submitted by Posiva to the Finnish authorities for a decision in principle according to the *Nuclear Energy Act*.

Reflecting the cross-section of stakeholders at each table, a number of different insights were introduced and discussed. At the end of the discussion, a facilitator for each table summarised the discussion and introduced the summaries in a short presentation. In the perspective of sharing the output among the tables, the summary was prepared in the form of a response to four pre-set questions as follows:

- Was the stakeholder involvement process sufficient?

- Did you receive all the information you needed for your involvement?

- What are the lessons learnt?

- How could your involvement be improved in the future?

As is evident, these questions were aimed primarily at the Finnish stakeholders at each table and, at most of the tables, their views were introduced briefly prior to the roundtable discussion. Since each table did not necessarily have an entire spectrum of stakeholders, the information provided and impressions perceived at the beginning differed from table to table.

Overall, deliberations for each question can be summarised as follows.

2.   Was the stakeholder involvement process sufficient?

From the implementor's perspective, there is no reason to disqualify the process and outcome of the EIA since it was completed as stipulated by the *Nuclear Energy Act*. It is also reasonable to say that the sufficiency of the EIA for the legislator was confirmed by the fact that the decision in principle was promulgated. Taken as a whole, the EIA thus appears to have been sufficiently dimensioned. It emerged from the discussion that a majority of the Finnish stakeholders present at the workshop were generally in favour of the EIA as a tool for involving stakeholders in the decision-making process.

However, regarding the details of the EIA process, there are comprehensive questions and concerns that can only be answered and valued by the parties directly involved. Among six plenary speakers in the session, there were two representatives from relevant national authorities, two representatives from research organisations, one representative from a concerned opposition group and one representative from the implementing organisation. For those who were not involved in nominating and inviting plenary speakers, it was not possible to determine whether this proportion represent the Finnish society at large or more importantly, the local community at the very sites in question.

A majority of the FSC members at each table shared the view that the EIA provided opportunities for those who have interests or concerns to become involved to an appropriate degree. One concerned stakeholder, on the other hand, clearly stated that the EIA was a "theatre" for all players to act out the scenario prepared for achieving the pre-set goal of acceptance.

The depth as well as the widths of discussion on this topic seemed to vary between the tables mainly reflecting the different perspective of Finnish stakeholders at each table. According to the summaries presented, the possibly closed character of the EIA (or other parts of the decision process) did not dominate discussion.

Last but not least, it should be noted that the participation of STUK (the safety authority) from an early stage of the EIA was highly valued. The independence and technical competence of STUK appeared to be acknowledged by a majority of the stakeholders. In this way, STUK was able to play the role of guarantor of safety, freeing stakeholders to delve into other aspects of the project at issue

3.   Did you receive all the information you needed for your involvement?

Among the Finnish stakeholders around the table, there were both providers and receivers of information. In general, a majority believed that there was sufficient information available. Some even claimed that there was too much information.

The session moderator noted it is not only the volume of information that enhances the credibility of the programme; equally important is providing appropriate information in a timely manner with a sufficient level of detail.

In this connection, a majority of the participants deeply shared a view as one of the Posiva representatives touched upon the importance of continued daily dialogues with local people to build up a firm basis for more dedicated information and discussions to follow. Suggestions were made regarding ways in which the contact person for the EIA at each site might have played a more efficient outreach role.

Some negative reactions to Posiva's newspaper advertising campaign perceived as scornful of public beliefs were recalled. There were some comments that the quality and quantity of information about alternative measures for disposal was not sufficient to allow a comparative study of disposal options. An observation was also made on the lack of resources for the opponents to develop their own information. Another question remained as to whether sufficient information had been provided about relevant projects in other countries where requested.

4.   What are the lessons learnt?

In a good tradition of the FSC activities, there had been active discussions among the members of FSC core group and the NEA secretariat prior to the Workshop. This preparation was quite helpful for the moderator of the discussion. Some of the FSC representatives from outside Finland noted that English-language web site had provided a great aid for them to get acquainted with or reconfirm their knowledge of the Finnish programme in preparing for the workshop.

The Workshop provided unique opportunities to confirm the background as well as history leading up to the decision in principle, in particular with regard to the EIA process in Session III. For a majority of the FSC representatives from outside Finland, the face to face dialogue with a spectrum (if not the comprehensive range) of Finnish stakeholders provided great opportunities to hear unfiltered voices something not easy to achieve through one-way information relying on printed matter or web sites.

For those who attended from abroad, it can be said that the lessons were learnt by "first-hand impression" rather than by formal and detailed information only, which alone made the trip to Finland rewarding. A majority of the participants have learnt a lot from the Workshop as summarised below. There are some remarks contradictory to each other. Such remarks are introduced in parallel since it is not an intention of the Workshop either to judge or prioritise the different views specific to a country.

- The overall waste disposal programme and its time schedule, including the EIA processes were consolidated based on a national legislation. Accordingly, the role and involvement of stakeholders in the EIA process were open and clear from an early stage. Although there was opposition to the "pre-set scenario" of the EIA process, a majority of the stakeholders attended to the workshop support the EIA as a multifunctional tool.

- There have been a voice that it is beneficial to allow the stakeholders to participate in a more informed manner during the early stages of EIA process, with the potential to resolve at least some of their concerns. It is less efficient solely to use the more formal consultative process near the decision point.

- The extent of stakeholder involvement in the EIA process is dependent and the level of interest and willingness to participate on the part of those affected by the project. As the Finnish example shows, the role of the EIA in the decision-making process for the project is fundamental to the willingness of the stakeholders to be involved.

- There has been an opinion that the EIA process was a purpose built (goal oriented) performance with a pre-set scenario in which the players are acting there given roles.

- Interestingly, the public participation level decreased throughout the process. The public apathy may be due to some extent on a failure to realise the relevance of invitation to become involved. The relationship between the EIA and the decision-making process was not fully clear at least for the general public.

- The objective of stakeholder participation within the EIA process should be made clear to the public before the process begins. The role of the EIA within the decision-making process is also crucial, as, if the EIA has little weight, it may be difficult to persuade the public of the value of participating. Here, local and national governments have some responsibility for encouraging understanding of opportunities amongst both the general public and developers.

- Special emphasis seems to be placed on Finnish society upon introducing realistic alternatives wherever possible in order to facilitate constructive discussions in decision-making processes.

- The recognition of, and trust in, STUK (the safety authority) among a majority of the stakeholders as a highly competent party is significant in comparison with what is frequently seen in other cases in the world.

- As has been said many times, Posiva's continued efforts to maintain everyday dialogues with the local stakeholders played a crucial role in paving the way for the EIA process.

The Finnish stakeholders at each table participated actively and fulfilled their role of informing visitors to a high level, regardless of their personal perspectives. Some of them reflected, in hindsight, upon problems that included the confusion between the EIA and the decision in principle, whether it was appropriate for the Ministry of Trade and Industry (MTI) instead of an environmental authorities to handle the issues, lack of information about the alternatives to geological disposal and so on. However, given the time limitations, these issues were not discussed in depth during the session.

5.   How could your involvement be improved in the future?

This question was addressed mainly to the Finnish stakeholders with the aim of further improving their involvement in the future. From the FSC members' perspective, it provided inspiring suggestions for those who are facing, or will soon face, similar phases of stakeholder involvement. The discussion around each table included the following recommendations:

- The complexity of the EIA process should be simplified. Public participation should be made as simple and accessible as possible. This will encourage wider involvement in the process.

- Sufficient information should be provided on the alternatives to facilitate decision making.

- It is always important to listen to other stakeholders and/or the public and to provide answers to their concerns during the EIA.

- A way should be found to distribute the proper level of resources each stakeholder requires for confidence in the ability of the EIA process to capture all points of view.

Moderator's remark

Beyond the horizon of the Workshop, the moderator observed something that he believes to be important to bear in mind when learning lessons from the Finnish experiences. The point to be noted is the extremely severe natural conditions in winter that, for generations, Finnish people have had to live with. Regardless of their position on nuclear issues, they are aware of what their quality of life would be if their energy supply declined, if not ceased, for any reason. For the Finnish people, and for Finland as a nation, may it possibly seem like a simple question of "do or die" in an immediate manner than most of the others. As the moderator has had the experience of living in an arctic region in winter with a relatively uncivilised infrastructure, he can share the views of the Finnish people on energy. It is a living condition where the refrigerator is expected to keep water in liquid state. When the FSC members visited the Olkiluoto power plant site, they learned that the net working rate of the Finnish reactors is very high compared to the average. The moderator considers that, if Finland were located in southern Europe, they might not have reached the stage that the rest of the nuclear world currently admires.

SESSION IV

CONFIDENCE BUILDING: WHAT GIVES CONFIDENCE
TO THE VARIOUS CATEGORIES OF STAKEHOLDERS?

Chair: J. Lang-Lenton
ENRESA, Spain
Moderator: S. Webster
European Commission

WHAT GIVES CONFIDENCE TO THE VARIOUS CATEGORIES OF STAKEHOLDERS?

J. Andersson
Member of the Green Parliamentary Group, Finland

Green members of Parliament were actively supporting the change to the *Nuclear Energy Act* in 1987 when export of the spent fuel radioactive waste was banned. After that all spent nuclear fuel produced by the Finnish nuclear power plants has been disposed and stored in Finland. The Greens think that this decision is still the only right one although, we are now facing the situation where we have to decide the fate of the spent nuclear fuel.

Parliament ratified the Decision in Principle on the final disposal facility for spent nuclear fuel in Olkiluoto in May this year. The Greens started the preparations of the decision making by arranging a fact finding excursion to Posiva's investigation site in Olkiluoto. The excursion was hosted by Posiva Oy. In addition to this, Greens arranged an open seminar on disposal of spent nuclear fuel. The aim of the seminar was to hear authorities and experts many-sidedly around the issue. As background and support for the decision making have Greens got information from many various sources like for example from Ministry of Trade and Industry, Ministry of Environment, Posiva Oy, STUK, OECD and environmental organisations (NGO). Let me mention, that open information strategy that Posiva has chosen in lobbying the decision makers has received recognition among the Greens. On the other hand the commercial blitz which was launched by Posiva in order to influence the common opinion was to our opinion tasteless.

Parliament's ratification was a very important phase at the process. The Finnish Parliament is not able to directly affect to the future of the disposal of spent nuclear fuel according to the *Nuclear Energy Act* after the Decision in Principle. During the decision-making process in Parliament in its committees our MPs were able to hear amounts of authorities and experts. Anyhow, we think that the decision making over the final disposal of spent nuclear fuel went all too fast compared to what is its significance from for example the global and temporal perspective. No other country has so far gone as far as Finland in its decision making over disposal of spent nuclear fuel. The non-governmental environmental organisations criticised Parliament because their voice was not heard extensively enough in the process. As a whole, there was all to short a time given to the experts in the committee hearing compared to the similar decision-making process in Britain according to the NGOs.

Although it seems like that different parties or stakeholders would have had a real possibility to influence the Greens in the decision-making process our trust towards the final disposal of spent fuel has not been growing. We consider the term "final disposal". For the Greens the ratification of the Decision in Principle does not mean that the disposal of the spent nuclear fuel should be finally solved.

The Greens have a qualified position towards the disposal plan. There are still many question marks in the plan which have to be investigated more closely. For example, to our opinion Posiva should give attention to the positions of the Environmental Committee. According to its position there should be made careful pre- investigation of the rock foundation from the surface before construction of an underground shaft. The results of the pre-investigation should also be evaluated carefully and openly. This is mentioned because Posiva's position towards the need and extend of the pre-investigations and its timetable has not come clear to us: In the application for the Decision in Principle there is said that the constructing of the shaft is going to be started already in 2003-2004. However, in the committee hearings in Parliament and after that there have come forward circumstances (e.g., Geological Survey of Finland and STUK) which refer that Posiva should not start the constructing of the shaft before careful pre-investigation of the rock foundation, which will take 5-10 years.

There has been mentioned our concern about the pre-investigation of the rock foundation. As much as that we are waiting for the investigation results concerning the heat of the nuclear canisters, water flow, the salinity of the ground water on the planned disposal depth as well as the effects of the glaciation to the nuclear waste canisters and to the bentonite before giving our support to the plan.

Even though our attitude towards the plan is qualified are we strongly supporting further investigation of the final disposal possibilities. The only way to increase our confidence is to make the investigations carefully, using the best available technology and evaluating the results openly and as broadly-based as possibly (also internationally). Our future attitudes will be influenced by the way how the investigation results affect the disposal plans: For example whether there is readiness to abandon some of the choices which have been made before the investigations and willingness to present new solutions for the disposal.

We hope that the "forerunner" role does not make blind but there is readiness to slow down the speed of the progression if necessary. If for example the pre-investigation of the rock foundation shows reason for it. The authorities supervising the plan have the possibility to make new decisions concerning the principle of the nuclear disposal according to what comes up in the research. This possibility must also take into use if necessary. In that sense the Decision in Principle is to our opinion an interim decision as its nature.

Parliament included a statement in the Decision in Principle. According to that statement government has to table a brief before Parliament concerning the new investigation information and technical development gained after making the Decision in Principle to be able to make sure the security aspect of the final disposal. We expect that the clearance is going to be profound and that Parliament gets the opportunity to consider the final disposal solution on the basis of it.

CONFIDENCE BUILDING:
WHAT GIVES CONFIDENCE TO THE VARIOUS CATEGORIES OF STAKEHOLDERS?

A. Lucander
Member of Eurajoki Municipality Council, Finland

During the fuelling of the first Olkiluoto Nuclear Power Plant Unit, discussions about sending spent fuel abroad for reprocessing were very topical and the people of Eurajoki got the impression that spent fuel will not be finally disposed of at Eurajoki. This principle was even written down in the long term strategy documents of Eurajoki and it remained there even for a long time after the government of Finland in 1983 had requested TVO to prepare a time schedule and plans for the final disposal of spent fuel in Finland. It did state that the municipality shall act so that high-active spent fuel will not be disposed of in the Eurajoki area.

While TVO accordingly investigated the national solution for final disposal, spent nuclear fuel from the Loviisa nuclear power plant was still returned to Soviet Union. This inspired to a certain extent lack of confidence within the inhabitants of Eurajoki and quite many of them had the feeling that TVO had broken its promise. Those who were especially critical towards nuclear energy in general used the argument that spent fuel was promised to be sent abroad for reprocessing and only the residual high-active waste would come back, if any.

So this topic has been discussed at Eurajoki generally and in the municipal council especially both in connection with final disposal plans and also in connection with operating licence renewals, power upgrade procedures and Decision in Principle Applications during almost 20 years. Good for confidence building is that the information policy of the utility has been very open, which has also been requested by the safety authority STUK. All incidents in the plants have been reported without delay to authorities and also given to publicity even when not safety related. While operating results of both Olkiluoto units have been of internationally top class this has contributed to the formation of confidence.

Routines for dialogue between municipality and utilities have been created in order to guarantee continuity. Thus several types of liaison groups were formed like:

- Liaison Group between TVO and Eurajoki with its neighbour municipalities for general nuclear power information topics.

- Liaison group between TVO and Eurajoki for nuclear waste information topics. After founding of Posiva Oy it continued with Eurajoki.

- Liaison group between TVO and Eurajoki for more detailed dialogue.

Also there has been many meetings for information and dialogue between TVO and respectively Posiva Oy with the whole municipal council and board on topical issues.

About 160 of the roughly 5 900 inhabitants of Eurajoki work at Olkiluoto. Even though it is a small share, it is however a good share to spread information and to build up confidence especially as the operation experiences of both Olkiluoto Nuclear Power Plant Units have been excellent even internationally. So confidence building has in our case been a long process.

While the more detailed site investigations started in 1992 revealed that bedrock at Olkiluoto might be suitable for a final depository, discussions started in the municipal council about removal of the prohibiting clause in our long-term strategy document. Motivation for this was to start the dialogue about the possible location from a clean table so that the answer will not be given before the question has been presented. The clause stayed in voting 1993 with votes 13-13 (one abstained), however, next year it was deleted with votes 15-10 (2 abstained). The council members were exactly the same, but earlier that same year the Finnish government had prohibited export of spent fuel. At the same time, import of radioactive waste was prohibited by law. A good share of people had previously felt the risk that spent fuel would be imported from other countries. It still remained as argument for those generally opposing nuclear power. Consequently, a possibility of foreign material import was always strictly excluded. However, when it came out that spent fuel of the Loviisa plant cannot be sent to the Soviet Union, it was understood that fuel from Loviisa would be put to same repository. This did not complicate the dialogue at Eurajoki, except for the part of transport.

Also the same year an opinion survey was made at three investigation sites. At Eurajoki the results showed that about 40 percent of people would accept final disposal at Eurajoki, if the bedrock investigations would give positive results. Opinion surveys also revealed what type of information was needed for different groups of people.

The dialogue between Posiva Oy and people of Eurajoki preceding the Decision in Principle has been long and thorough. The EIA-process formed an excellent platform for dialogue where interested people could participate with their critics, questions and requests for additional clarification. Similarly, the municipality could comment on the research programme parts of topical interest for final decision making like social, health, economical, employment and image impact evaluations etc.

In 1997, Eurajoki hired a consultant to perform a competitiveness analysis to chart our relative strengths and weaknesses compared with other municipalities. Following this in 1998, the municipality and local entrepreneur association performed together a SWOT/four-field analysis charting our strengths, weaknesses, opportunities and threats. We then compared "scenario of possibilities" with "catastrophe scenario" and developed our main strategy of Eurajoki 2000. This strategy included the so-called "Olkiluoto vision". In that vision, a spent fuel repository was included. All these scenarios and visions were openly discussed and accepted by the municipality council. Local press has taken good care of publicity.

Having these processes going we were well prepared to participate in the dialog around the EIA-programme. In that dialog, many local resident groups were represented in addition to the municipality organisation. Also neighbour municipalities participated in that dialogue. Local and provincial nature, health and environmental organisations etc. were asked to participate as well as any private persons. A neutral opinion survey was made as a part of EIA programme. It revealed that, in 2000, 59% of the population were willing to accept a repository at Eurajoki if the local bedrock proves to be good enough for it.

So in its meeting on 24 January 2000 the Municipality Council accepted Posiva's Application-in-Principle for a final repository. As a proof that wide confidence had been built was that Parliament accepted the application this year with a large majority, only three Parliament Members were against.

SOCIAL SCIENCE AND NUCLEAR WASTE MANAGEMENT: BUILDING CONFIDENCE, LEGITIMISING DECISIONS OR INCREASING COMPREHENSION?

T. Litmanen
Senior Lecturer, University of Jyväskylä, Finland

Abstract

The societal evaluation of the risks associated with nuclear technology has now continued for over fifty years. Different actors, such as anti-nuclear movements and the nuclear industry, have tried to establish a dominant position in the debate. However, despite the vast amount of research done in technical and natural sciences, debate over the issue continues, and there is not consensus in sight. For the opponents of nuclear technology, the risks are much too high to accept, and for the proponents the risks are controllable and thus at an acceptable level.

For a sociologist, it is a challenging task to study this kind of a controversy where each party claims they are absolutely right and that they are the only ones who hold the truth about an issue. What does the disagreement indicate? Why have not the parties found a solution, or a reasonable compromise? Which of them holds the truth? Is there a truth in this issue? By focusing on these kinds of questions, the presentations discuss the role of social science in Finnish nuclear waste management system? It will give a short overview of the way social science was integrated to the process of developing and planning waste management and describe the issues, which have studied in public sector's research programme. It will be argued that in building confidence an important task is to increase overall comprehension of the parties through applied social science research but also through profound theoretical research.

THE ROLE OF THE PUBLIC SECTOR'S RESEARCH PROGRAMME IN SUPPORT OF THE AUTHORITIES AND IN BUILDING CONFIDENCE ON THE SAFETY OF SPENT FUEL DISPOSAL

S. Vuori
Research Manager, VTT Energy
K. Rasilainen
VTT Energy

Summary

A multiphase research programme was launched in 1989 to support the Finnish authorities in their activities concerning spent fuel management. The Finnish programme for spent fuel management has so far managed to keep its original time schedule at least partly due to clearly defined responsibilities between the nuclear energy producing industry and the authorities. It appears that the public sector's research programme has been successful in its supporting role by providing research results both on technical/ natural science and social science issues. In addition, the research programme has contributed directly and indirectly in building confidence on the post-closure and operational safety of a spent fuel disposal facility.

The main objective of the multistage Public Sector's Nuclear Waste Management Research Programme (JYT) has been to provide the authorities with expertise and research results relevant for the safety of nuclear waste management to support the various activities of the authorities. The main emphasis in this multidisciplinary research has been paid to the final disposal of spent fuel.

The first phase (1989-1993) of the research programme was traditional technology and natural science oriented research, but during the second phase (1994-1996) socio-political and societal issues became also part of the programme. In the third phase (1997-2001), these social science issues have become quite central themes. This is understandable, because the implementation of a technical plan to handle spent fuel requires that the plan is accepted more broadly than solely within the circle of nuclear waste management experts. Practical decision making about spent fuel management has proved to be difficult in many countries. This is at least partly because there are also other arguments than technological and scientific ones. Furthermore, there are also other stakeholders involved than the implementor and regulator.

Due to the small amount of resources available in Finland, the public sector's research programme has not aimed to do independent full-scale performance assessments of spent fuel repository. Rather, emphasis has been placed on studies to reduce uncertainties associated with the basic principles and main phenomena related to the geological disposal of spent fuel, and to be able to model these factors more accurately and reliably for safety assessment purposes. In addition, state-of-the-art reviews on selected key safety topics have been prepared to support the review work of the safety authority (STUK). The latest major reviews have covered bedrock stability, the prospects of coupled modelling and the retardation mechanisms. One major area in the research programme covered the natural analogue studies at the Palmottu U deposit in southern Finland both as a domestic in situ migration research project, and later as a broader scope EU project. This research topic has brought better understanding on some key phenomena relevant for long-term safety. At the same time it has provided an alternative way to illustrate the safety function of natural barriers included in the spent fuel disposal concept.

The role of the public sector's research programme was not, however, restricted to the technical support services to the authorities. For the decision making in local and national level the Environmental Impact Assessment (EIA) and Decision in Principle (DiP) processes carried out in the period 1998 to 2001 were the most decisive phases. The background material provided by Posiva was quite extensive and, therefore, there was a real need for concise and impartial support documentation for the use in local and national decision making. A set of three reports was prepared within the JYT-research programme covering the following themes: Main features of the Finnish plan for spent nuclear fuel management; the principles of safety assessment in spent fuel management; and the illustration of the radiological impacts evaluated in the safety assessments for the spent fuel disposal facility and the transportation of spent fuel. Furthermore, a report on the status and feasibility of alternative solutions (i.e., partitioning & transmutation and very long-term interim storage) was prepared for the use of authorities in decision making related to EIA- and DiP-processes.

The implementing organisation, Posiva Oy, has had the main responsibility to conduct research also on societal issues in the context of the environmental impact assessment (EIA) procedure. However, increased confidence among the public in the affected candidate municipalities has probably been achieved by the complementary studies conducted within the public sector's research programme on topics related to environmental impacts, social impacts and image issues. Furthermore, in the Environmental Impact Assessment (EIA) procedure pertinent projects of the public sector's research programme provided substantial support to KTM in the follow-up of the EIA procedure. Furthermore, the projects gave statements on the substance of the EIA report by Posiva.

One general feature facilitating the building of confidence on the Finnish nuclear waste management programme has been the clear, long-term schedule defined by the government's policy decision in 1983. In the stepwise process the authorities have had possibilities to follow-up the development of disposal plans and to conduct at intermediate milestones reviews concerning the post-closure safety. The DiP process is internationally unique and, therefore, not easy to communicate. While providing an absolute veto right to the proposed host municipality and requiring the final ratification by Parliament, the DiP process has added a considerable amount of commitment to the process from the part of political decision makers, on both local and national level. It reduces the possible press on the decision makers by the applicant, because no substantial investments in the facilities are allowed prior to the approval of the DiP. It is of help to the implementor in that the political decision makers have agreed with the fundamentals of the proposed plan at a sufficiently early stage. The EIA process was organised as a structural part of the DiP process. This gave the whole EIA process a clear technical meaning, something that undoubtedly helped in focusing the two processes.

SESSION IV REPORT
IMPORTANT ASPECTS IN CONFIDENCE BUILDING
(FINNISH CASE STUDY)

Moderator: S. Webster
European Commission

In no particular order of importance:

A clear government strategy for managing SNF, e.g., no export or import by law, and a clear recognition by the State of their moral responsibilities in this matter.

The legal framework, such as the important role given to the EIA process, especially as a platform for dialogue, the municipalities' right of veto and the modalities of the Decision in Principle.

Openness displayed by Posiva during the process.

Good track record of the operating Finnish NPPs and the confidence that this has fostered in the local communities.

The inclusion of retrievability in the debate, and the increased acceptance of this concept by the developers.

The research carried out in the public sector, of both a technical and societal nature, destined to aid the local and national decision makers as well as informing the public.

The fact that STUK are widely held as being technically competent and have maintained their independence and credibility despite a pro-active involvement in the process, involving their clear explanation of their role in the process.

The trust by the municipality of Eurajoki in the other stakeholders, such as STUK, Posiva and the Ministry of Trade and Industry, and also in the legal framework, especially the right of veto the decoupling of the DiP from the debate about the fifth NPP.

Certain cultural or traditional Finnish characteristics: faith in their democratic institutions; consensus-orientated approach; readiness to accept technological innovation

In general, confidence building is essential to ensure that the maximum number of stakeholders have faith that the process as a whole is capable of yielding a "fair" result. Even of the final decision is contrary to the views of one or more groups, providing these groups see the process as fair and transparent there is more likelihood that they will accept the outcome. Some concessions along the way (e.g., in the Finnish case, the inclusion of greater retrievability in the concept) also help ensure greater acceptance of the outcome, even if there is no agreement with the final result.

However, confidence and trust is very hard to earn (in the case of Finland, some of the above bullet points have taken many years – if not decades – to put into place) and very easy to lose.

1.   Ranking of the confidence-building measures

This is really a question for the Finnish stakeholders themselves, and it is likely that each group will have its own opinion regarding which measures were more or less important. Indeed, all of the points raised in Section 1 above can be considered as important. Considering the case of the municipality, it is evident that their right of veto, the clear government strategy regarding SNF and the role of public participation as determined by the EIA were crucial. Some workshop participants considered that the institutional measures were, in general, the most important, followed by the social and then technical measures, though technical measures would gain in dominance as the programme progresses. However, what is importance is to maintain the dialogue (which means *two-way* communication), since the process is a step-wise one and the trust and confidence that have already been generated need to be maintained throughout the whole duration of the project (i.e., design, construction, operation and closure!).

2.   Expressions of negative experiences

The points mentioned in Section 1 constitute the more positive experiences. As for the more negative ones, the following were noted. Most of these points were raised by members of the opponent (or, at least, sceptical) groups present at the workshop:

Posiva's "public outreach" publicity campaign was considered by many to be in bad taste.

The institutions that were responsible for carrying out the public sector research were considered by some not to be "neutral" in the nuclear debate.

Some thought the Ministry of Trade and Industry (which also regulates the nuclear sector activities) to be an inappropriate choice for competent authority in the case of the EIA, even though this was dictated by the Act on Nuclear Energy.

The wording of the questions asked in the public opinion surveys were not always considered to be fair or appropriate.

In the past there had been a change of policy on the part of the government regarding export of SNF, and this could reduce the confidence in the government's present policy. The fact that the decree to halt exports of VVER SNF came into effect only in 1996 was not considered by some to be respecting the spirit of the decision taken by the government[13] in 1994.

Some people felt that the process in Finland was going too fast and there was no advantage in being the first country to decide on the siting of a SNF/HLW repository.

---

13. In 1994, Parliament made an amendment to *Nuclear Energy Act*, where export and import of nuclear waste was prohibited. The decree prohibiting the export of nuclear waste came, however, into force from the beginning of November 1996. This was due the agreement of shipment of nuclear waste between IVO and Russians – this agreement was still valid when the amendment was taken and had to be finalised.

Doubt was also expressed over the policy not to allow Parliament any further role in the decision-making process following the endorsement of the government's Decision in Principle.

STUK may not have had in-depth technical review capability.

3. Lessons learnt? Possible improvements?

There are many things that can be learnt and inferred from the Finnish experience to date regarding confidence building and the gaining of trust. Most of these are, perhaps, just common sense, but nonetheless it is important to note them:

- Be open and honest!

- Involve stakeholders early and interact with them often.

- Try to bring into the debate people who may have different views.

- Adopt a stepwise approach to decision making and ensure that the strategy and procedures are understood and accepted by as many people as possible.

However, there was a general feeling that the process was not over yet and that safety still needs to be proven. The MP present at the workshop was quick to point out that Parliament had only given its approval for a programme of investigation at the site, not for the construction of an actual repository. This emphasises the importance of stepwise decision making, and the crucial role that will eventually be played by STUK. Trust and credibility still need to be maintained, especially by Posiva, during the coming years, and Posiva should remain attentive to public concerns as far as possible and seek to address them at every opportunity.

In the context of the moral obligations of the present generation, it seems to be generally accepted that geological disposal of SNF/HLW is better than leaving the waste on the surface. However, the importance of retrievability in the eyes of the public should not be underestimated.

The role and importance of EIA was probably more significant than had been anticipated at the start of the decision-making process, not least because of the mandatory involvement of the public. It is clear that this is one of the key mechanisms to assist confidence building in any project of a controversial nature with potentially significant environmental impacts.

Finally, a word about the applicability of these lessons to the situation in other countries. There was considerable uncertainty expressed by many people as to whether the Finnish case is indicative of what can be achieved in different cultural, social, democratic and project-development contexts. It must be remembered that some of the prerequisites for the successful Finnish DiP were laid down many years ago, perhaps even decades or centuries in the case of those of a more cultural nature. It is therefore difficult to imagine that these conditions could readily be reproduced elsewhere. Nonetheless, any national disposal programme undertaken in a Western democratic country would probably benefit from an in-depth appreciation of the practices in and the lessons learnt during the Finnish programme.

SESSION V

CONCLUSIONS, ASSESSMENT AND FEEDBACK

Chair: Y. Le Bars
President, Andra, France
and
Chairman of the Forum on Stakeholder Confidence

THEMATIC REPORT ON PUBLIC GOVERNANCE

F. Bouder
OECD/PUMA

The objective of the FSC workshop was to learn from the process that led to the ratification by the Finnish Parliament, on 18 May 2001, of the Decision in Principle on the final disposal facility for spent nuclear fuel in Olkiluoto, Eurajoki. One of the objectives clearly expressed by the NEA was to help government to make the right decisions, including by looking at successful policy making and decision-making practices. This gave "Public Governance"[14] issues a prominent role in the debates. In addition, the meeting was conceived as a discussion between stakeholders, which, interestingly, also made it an exercise in "Public Governance".

This purpose of this report is to identify some key governance lessons that emerged from looking at this process in detail. It looks at the difficult policy context in which the Decision in Principle had to be taken. It highlights the governance challenge that emerged and suggests some key findings that could possibly be relevant in other similar circumstances.

1. The policy context

The policy context in which the Decision in Principle was taken is characterised, from a public management perspective, by the existence of a very challenging environment. The debate has been partly influenced by a mix of safety and social uncertainties, by the existence of complex institutional and procedural arrangements and by the fact that several interests would be concerned. This complex contextual reality certainly challenged the process for identifying a agreed sustainable solution for the site selection.

1.1 The risks

Nuclear Safety issues appeared to be a major area of public concern. This is not surprising, since Public opinion is generally very sensitive to potential risks when talking about a new nuclear installation. This concern is usually motivated by more general concerns relating to the use of nuclear energy. A usual complaint from scientists was that the site selection was supposed to be a specific decision but was "misunderstood" by people who considered it as a general decision on waste management. However, the workshop clearly underlined the limits of any "experts"[15] attempts to

---

14. We suggest, as an operational definition, to consider Public governance as the set of mechanisms, procedures and practices for managing the public sector effectively and ensuring that government is fully accountable to the rest of society

15. It is very difficult to define precisely the concept of expert. In the present context it cover both scientists and the main institutional actors present in the field including from the energy provider, the regulator, the bureaucracy etc.

totally separate the debate on general nuclear issues and the specific discussion on the identification of a specific site for the disposal of spent nuclear fuel.

The workshop provided an illustration of key aspects of risks perception by the public. Risk perception is based on the interpretation of certain "signals" analysed by human intelligence on the basis of existing evidence, past experiences and expected prejudices. Therefore it cannot be expected to be a "scientific" assessment. "Human" and "scientific" rationality are therefore different and this difference needs to be fully understood by the different actors of the policy debate. Therefore, in addition to technological risks, there are clearly "social" risks that should be taken into consideration with equal importance. The Workshop showed that Nuclear matters have often been perceived by many categories of the public with some degree of caution, scepticism or opposition resulting mainly from uncertainties relating to the exact degree of reliability of the nuclear technology and was not only driven by technical concerns on the degree of reliability of specific installation. The fact that the nuclear technology has no military applications in Finland may have contributed to limit some of the social risks.

Therefore it underlined that the risks about specific safety requirements of a specific site are only one part of the total picture and that social acceptance is also a major challenge and a source of potential risks of policy failure. The need to balance safety requirements with requirements "social acceptance" therefore appeared very clearly for the evaluation of potentials risks and obstacles.

These concerns are often expressed in the context of specific decisions, e.g., will the new installation meet all the expected safety requirements. However, longer-term concerns may also be significant, in particular concerning the existence of longer-term uncertainties about the possibility to eliminate waste (e.g., how will our choice impact on future generations). Therefore the time factor was recognised to be a particularly important parameter in the decision-making process.

Some criticisms from those opposed to the project were expressed at the meeting, showing concrete challenges to social consensus building. These criticisms, included the idea that the selection process is often laid by political and economical approaches rather than "technical", that acceptance is related to the image of the community where the installation will be located and not by "rational" criteria, and that the election process may jeopardise the selection process of a possible site.

The existence of initial suspicion from environmental organisations that "authorities" – including the Finnish expert organisation established by two nuclear power companies (TVO and IVO) in 1996 and responsible for the characterisation of sites for final disposal of spent fuel (Posiva Oy) – would not be totally neutral or would try to minimise some risks, illustrates the potential tensions between various actors and the importance of addressing the social risks.

1.2   The stakeholders

One of the fundamental complexities of the decision-making system was related to the significant number of different players. In addition to national and local government, to the scientific experts, to the regulator and the energy provider, various stakeholders with scientific and no scientific background could have been affected by the decision-making process. Environmental organisations, private business, unions, but also "ordinary citizens" (*laymen*) were concerned by the decision.

One of the consequences of the number of different players motivated by different perceptions, opinions, economic interests etc, is that virtually all groups and individuals around the site could develop potentially conflicting approaches and views. The site selection process of course could not have ignored this situation without a serious risk of jeopardising social support. One of the key questions for managing this complexity was to reach the kind of "morally sustainable solutions" that would generate social consensus.

The workshop highlighted some key elements on the main challenges that decision-makers should bear in mind before embarking in processes aiming to improve stakeholders involvement:

- Stakeholders have different perspectives and evaluate issues and risks according to their own concerns.

- The context matters (culture, background, given interests, etc.).

- Information, communication, consultation and participation are not the same and therefore the modalities of stakeholders involvement should be clearly identified according to their objectives.

1.3   *The political culture of Finland*

One of the main challenges when setting up the conditions of the site selection process was to design a process that would be consistent with the policy making culture of Finland. The Nordic model of government is characterised by strong democratic parties, a moderate "working multiparty" system and a consensual rather than adversarial decision-making culture. It operates under a system of extensive consultation, with a centralised system of bargaining. In short, the system has been characterised by compromise, co-operation and consensus.

In Finland, however this model is tempered by the dominance of the State and a more "authoritarian" strain in the culture. Some participants at the workshop stressed, for example, the high regard and obedience to the law. For example, it was highlighted that compared to other countries social movements appear to be less autonomous, more peacefulness, and demonstrations "have never turned into carnivals".

One of the key concerns for assigning responsibilities to various actors for the site selection process and for the elaborating the framework of the stakeholders involvement process is therefore certainly to develop rules and practices that do not contradict the Finnish decision-making culture. In particular, drawing lines of responsibilities that respond to the basic conditions of the policy-making culture of Finland has been recognised as an essential component of sound governance. It is particularly important to develop rather consensus-oriented mechanisms that will assure trust in the policy-making process.

The process that has been chosen met these concerns: For example it was decided that the Decision in Principle on the site selection should be ratified by parliament. The last vote was rather consensual. But it was also recognised that parliamentary decision should also be supported by sound interaction with society.

2.   Main governance challenges

This policy context created a very demanding situation in particular due to the political culture of Finland that is an incentive for more direct involvement of stakeholders than in many other countries. However the fact that there was a high level of trust in public institutions should not be neglected as a potential factor for success. The concerns for a safe and secure process created of course a challenging situation in particular with respect to the political acceptance among the general public, which is required by a consensus oriented culture like Finland.

2.1   *The procedure: adaptation and constraints*

The policy context in which the DIP took place was characterised by a mix of pre-existing procedures, in particular to ensure safety requirements, and by new provisions to improve stakeholders' involvement. The fact that the process for stakeholders' involvement was progressively formalised and used clear guidelines, was of much significance.

As in other countries the initial process was not so open to consultation; and the early stages of the site selection process (in particular the site identification and the site characterisation), were made without really involving stakeholders. Officials attributed this rather closed process to the technicality of the early steps in the selection process, or to the fact that it created a new policy debate with no real stakeholders groups to consult with. Whatever the reason for this, the initial policy was mainly conceived within Trade and Industry Ministry. However the idea that the debate was more than purely "technical" and should be brought to parliament was progressively recognised. For instance, Parliament debated on a proposal for a new power plant. Following a new, more inclusive process, the 2001 Nuclear waste facility government proposal could finally be endorsed by Parliament.

The full procedure, followed three main steps:

- Site identification (1983-85).

- Preliminary site characterisation (1986-92).

- Detailed site characterisation (1993-2000).

As mentioned above, the process ended by Parliament ratification in May 2001. It involved an adaptation of laws and regulations. During the third step contacts with stakeholders were organised (in 1997 and 1998): wide range of scientific experts, local government officials in host and neighbouring municipalities, national decision makers, regulators, environment organisations including opponent groups, and the media. The necessity to ensure compliance with a range of pre-existing procedures and rules throughout the site identification process was important. One could argue that procedural safeguards, although an essential component of a sound decision-making process, are also potentially a source of possible rigidities when they are not fully adapted to new conditions.

One obvious example is given by the necessity to conform with already formalised procedures of Environmental Impact Analysis (EIA). The impact of this procedural "constraints" on confidence is not easy to evaluate. EIA has sometimes been presented as an opportunity for building confidence. For some participants, however, EIA was perceived as an eminently "political" instrument and could therefore be problematic as an instrument of citizens involvement and acceptance building. For others EIA is seen as a technocratic tool, far too rational and logic to fit the requirements of the new, more open, processes. The EIA procedure is not a creation of civil society and is an institutional mechanism that provides, under the present circumstances, only limited room for active participation by the citizens. The procedure, as it is now, is carried through public hearings, written comments and an EIA contact person in each municipality. This, applied to the site selection process, created concerns among some stakeholders, in particular environmental organisations, based on the lack of room for interaction between the average citizen and decision-makers. Some participants at the workshop expressed their regrets that, although "Citizens are expert of local questions" their expertise was not fully acknowledge through the existing process.

2.2 Drawing lines of responsibilities

One of the key feature for the selection of a new site was the fact that this decision had tremendous local consequences, and therefore justified a key role for local government. Siting of facilities was recognised to be a key competence of the municipalities. It seems that only a minority of stakeholders would have preferred a purely local process, which would have been a difficult option considering the national dimension of the debate. Selecting few possible sites included gaining support at the local level, but the concern was also high that the site selection should not be reduced to a confrontation of local interests, to an exercise in local politics.

The idea that the importance of the debate would also require to involve the national level came out very clearly, and Finland chose that the final decision would remain at the national level, after identifying few possible sites in agreement with local government and stakeholders. One of the characteristics of the Nordic Countries, including Finland, is that the central role of state for regulating many aspects of the societal life is generally recognised. Therefore stakeholders tend to recognise that the final word on the site selection should belong to the National Parliament.

It appeared also essential to clarify the role of the regulator. Although it is meant to represent the public interest, the regulator – similarly to the other actors of the policy – is not totally independent and could develop biased views. Some participants at the workshop noticed that in the present context a question of division of power still exists, i.e., the regulator remaining to closely involved with the interests that they have to regulate (debate on independent regulatory authorities)

Therefore neutrality requirements should be reinforced when opening the dialogue to external stakeholders. It would for example not be acceptable to most stakeholders that the regulator would initiate campaigns. However expecting the regulator to stay at a purely "technical" level could be partly an illusion. Clear lines of responsibilities would be essential. The regulator should for example be active for decoupling waste management from other issues, and it should concentrate on the safety issues.

2.3 Stakeholders' involvement practices: critical issues

But what should be the extent of consultation and participation? Indeed the workshop underlined the existence of different perceptions of what it would mean to involve stakeholders. A fundamental parameter to take into consideration is the necessity to managing different stand points. For example the perception of nuclear power is partly influenced by one's background and opinion. One key criteria for social acceptance is therefore to understand clearly how the problem is defined, by whom. The question of who sets the agenda of the consultation and participation processes is of critical importance, as well as how the specific mechanisms for stakeholders involvement are put in place.

For most participant it seemed clear that key challenge for stakeholders involvement is that it should not only provide information to people, but it should be part of a management framework including open consultation and participation mechanisms. In the present context Information policies would very much be about releasing material in order to "convince" or "educate" stakeholders.

Consultation would require a capacity to receive and translate messages from stakeholders, while participation implies a full commitment to engage in a two-way dialogue.

In the case of the site selection process for the disposal of spent nuclear fuel, the fact that process of citizens involvement was not fully open when consultation started had certainly a negative impact and required sustained effort to rebuild and maintain public trust in the second, more open phase, of the process. This initial shortcoming was mainly attributed by some stakeholders to the rigid frameworks in the Ministry of Trade and Industry.

Another critical issues is the need to promote the credibility of the process. Acceptance should be a goal deserving efforts and investment but it is by no mean a guarantee of success. What would happen if consensus is not reached? This issue should clearly remain in the mind of decision-makers before embarking in consultation and participation.

Different stakeholders may also consider different steps to be important in the involvement process and therefore procedures can influence very much the perception of the value of the consultation and participation processes. There are also some critical issues about the expectations of the different actors. The public expects a final decision, while government tends to stress that the decision has to remain revertible. One related issue is also about how much Emphasis is put on alternatives to the "most likely" solution. Very often the decision is not perceived as very open by the citizens, while decision-makers stress the need for well-defined options in order to ensure a focussed discussion and to provide the conditions necessary for the later implementation of the decision.

Main issues for citizen involvement practices

The difficulty to create stakeholders consensus led to recognise the importance of the following elements for stakeholders involvement. We suggest that decision makers or the drivers of the process of stakeholders involvement ask the following questions:

- Who is involved and at what stage?
- Who monitors the agenda?
- What tools are used (to inform/consult/participate)?
- How are messages translated?
- What feedback is given?
- How open and transparent is the process?

There are also technical barriers that are inherent to extensive consultation and participation mechanisms: The large amount of partly contradictory information, the long-term and overlapping processes that this involves should incline to remain realistic about the outcomes the process.

The government's commitment to follow an open process is critical but sometimes there was a feeling that it would be jeopardised if not led by equal access rules and supported by adequate funding. For example both the experts of Posiva and the Regulator (STUK) stressed their willingness to treat all actors equally. However one of the challenge treating stakeholder as equal partners is that they do not have by nature the same weight in the policy debate: for example big semi-public companies have of course much bigger capacities than individual "laymen" to use the existing information, to analyse it and to make decisions on this basis. Only limited attention was paid during the meeting to the problem of either representing certain "disadvantaged" groups or enhancing their equal access to the process; however some point were raised that there were situations where stakeholders were uneven "David and Goliath", but that for instanced Goliath always managed to win the confrontation. Posiva certainly played a role to open up. How well Posiva has managed to get the trust of the public is certainly a matter subject to further improvements, according to some stakeholders. One improvement envisaged would be to encourage public financing of the consultation processes in order to support both NGOs independence and their capacity to take part in consultation processes. Public financing should not, however, create a situation of dependence that would harm democratic processes.

2.4   *Managing information: The call for transparency*

Making decisions "behind closed doors", even on the basis of reliable and accepted "technical" information seems no longer possible in the Nuclear Field. The variety of stakeholders potentially affected and even more the controversial nature of any nuclear related decisions have create a call for effective transparency as well as a pressure to release information without restriction.

However one should not underestimate the obstacles to developing a sound and comprehensive transparency policy. The most important of these obstacles is the fact that nuclear issues are sensitive and that some information may have a confidential nature. An equally important obstacle to developing transparency policies is the difficulty to overcome misunderstanding when it comes to sharing a common language among stakeholders.

Therefore, participants stressed the importance of developing an open communication strategy. In particular it could be a promising to associate the media to the site selection process. Partnership with the media may help to overcome some obstacle to knowledge sharing. However designing a proactive communication strategy could also create a number of concerns. For example one should ensure that it does not take place at the expense of the freedom of information and the process should ensure the accuracy of the information released. Concrete obstacles to a clear and informed debate should not be underestimated and would certainly require a number of specific safeguards.

3.   Main findings of the workshop

The workshop provided the opportunity to develop some key findings from the Finnish experience. These findings included a number of requirements to ensure a fair involvement of stakeholders in the decision-making process. We would suggest that they could be used as a preliminary basis for developing a framework for stakeholders' involvement in the Nuclear field. The main criteria, which are mutually reinforcing with the findings of a recent OECD Publications,[16] would be the following:

- To maintain commitment over time at all levels.

- To adapt legal and regulatory frameworks on a regular basis.

- To review the distribution of responsibilities and of regulatory authority.

- To ensure accountability and transparency.

- To create room for real and honest dialogue.

---

16. OECD 2001, Citizens as Partners, OECD Paris and OECD 2001, Governance for Sustainable Development, OECD Paris

3.1   Maintaining commitment over time

One key lesson from the experience of stakeholders' involvement in Finland is that a commitment to engage citizens in the decision-making process should be maintained over time and at all levels. In particular, political commitment is important to create a sense of leadership, which is a key element of the success of the process. But government commitment is not enough. There is also an important need to have commitment of the different non-governmental actors.

This commitment should also be sustainable. Longer-term commitment is critical but one should not underestimate the challenging nature of developing consistent policies over time. This can be attributed to four to five years electoral cycles, to the nature of the role of politicians, to the existence of conflicting agendas at the various levels of decision making.

The process launched after 1993 for the detailed site characterisation provided a clear illustration of this issue: it was supported by political commitment, introduced a management and goal oriented process with key rules and responsibility lines, good funding levels, developed a reasonable and realistic schedule, delivered clear messages for the site selection (for example Posiva has not been looking any longer for "the best site" but for a "good enough one").

3.2   Reviewing the distribution of responsibilities and of regulatory authority

The distribution of power at the different stages of the decision-making process needs to be checked against. what should be the respective role of local authorities for the decision, the role of parliament and of independent authorities should be clearly stated and, considering the fact that this process is a longer-term project and involves many different actors, the criteria of consistency will become essential. A sound distributions of responsibilities between the different actors has proved to be a critical factor of success of the stakeholders' involvement. One key question is about who should be in charge of specific decisions at each stage of the decision-making process. In the case of the site selection process for the disposal of spent nuclear fuel, the fact that a mix of independent agencies, environmental authorities, local authorities and central government were involved certainly contributed to building support and creating a sense of a rational decision-making process.

The importance of the regulator was not extensively developed during the workshop although it represents of course a key aspect of the decision-making processes. The legal and regulatory frameworks have often been established at a time when decision-making mechanisms where not so open to stakeholders' involvement. Therefore some consideration should be given to adapting the practices in order to guarantee that procedural safeguards are in place and to ensure the neutrality of the regulatory decisions. In the context of a more open decision-making processes, the demands of stakeholders will be higher and citizens will be reluctant to accept decisions that they would considered to be partisan.

3.3   Ensuring accountability and transparency

The existence of accountability and transparency mechanisms is also essential. One argument often developed for restraining transparency, in addition to safety restrictions, is the fact that few people come actually to the meetings when a consultation is organised. The fact that only few people decided to participate actively in a consultation or participation process should not be a reason for restraining access to the process. Non-participation does not mean that stakeholders are not interested, passive or against the project. It does more often mean that the agenda for the discussion or its content are not clear enough or do not raise the right issues or should be made more relevant and more accurate.

The fact that the last word on the site selection is given to parliament and not to the executive was for example a strategic decision had certainly very positive implications in terms of ensuring accountability. One could maybe explore what additional reporting mechanisms could be developed.

The discussion did not make extensive reference to the role of the media, although it was generally recognised that it constitutes a key criteria for enhancing transparency. In particular the idea that media could help clarify the debate was put forward. Of course, as it was already highlighted, this should not take place at the expense of the freedom of information.

3.4 *Creating room for a honest dialogue*

The idea that credibility is a very precious asset that should be constantly reaffirmed was noticed as a very important factor. Shall the process of stakeholders' involvement be poorly managed, credibility could drop easily and would become then very difficult to regain.

A condition to maintain credibility is to engage in a dialogue that is perceived by all stakeholders as a fair process, whatever the outcome of the consultation should be. The process should not give impression that choices are made in advance. The attitude of Posiva, perceived as keen to listen to concerns of the various stakeholders rather than only "talking" and trying to "educate people" had a significant positive impact on the process.

Creating room for debate and honest dialogue also implies to provide feedback to participants through the consultation/participation process. Professional training should help employees of the different organisations involved to change their attitude vis-à-vis some of the most critical stakeholders. Some key criteria for a positive involvement of stakeholders include:

- Good planning of participation, should take all parties views seriously.
- Taking these arrangements into details.
- Creating impartial and competent "facilitators" with a neutral voice financially independent.
- Relevant information based on answering questions rather than "throwing out data".

4. Conclusion

The FSC Workshop certainly succeeded in advancing the agenda of stakeholders' involvement in the nuclear field. Although the meeting focussed on a specific experience, i.e., the process of decision making for the disposal of spent nuclear field in Finland, the effects of this multi-stakeholder dialogue are actually of much more significant relevance. This should be attributed, from a Public Governance point of view, to the Particular attention paid to the complex role of non-technical stakeholders. Since the beginning of the meeting, the need for balancing technical solutions and the requirements of a fully democratic process was recognised.

In the nuclear field where the degree of acceptance of decisions has been declining over the past three decades, the need to build confidence among citizens is being recognised. The usual answer given by "experts" to this challenging of decline of public trust was, in many other field, to call for more education of the general public to scientific realities. If people do not accept some key decisions is it because they have a lack of knowledge of what would be the consequences? Is it because they are only influenced by irrational fears coming from ignorance? The FSC Workshop showed that the answer to these questions is much more complex than this.

THEMATIC REPORT ON SOCIAL PSYCHOLOGY

C. Mays
Institut Symlog, France

1. Introduction

Social psychology is concerned with the interactions among people and groups, and with their gradual formation of shared – or conflicting – attitudes, opinions and understandings. In preparing to provide feedback to the FSC Turku Workshop, I anticipated that it would be useful to structure my observations around concepts like:

Social identity: our idea of ourselves as it is formed through our participation in groups or institutions, e.g., our various identities as stakeholders in the radioactive waste or spent fuel management process;

Role: the set of actions that are expected of us by the group or institution to which we belong, e.g., our various missions to be carried out in regard to radioactive waste management or in regard to attending the workshop;

Social representations: shared sets of beliefs and values that act as a framework for explaining and evaluating events, e.g., our various assumptions about what is right, or important, or wrong, or inconsequential, in the radioactive waste management debate;

Demand: the wishes and needs, linked to our social identity and role, that influenced our participation here as well as our expectations for others, e.g., our desire to understand the unique or generalisable features of the Finnish process;[17] and

Definition of the situation: our active construction and interpretation of events, which finds itself in tacit competition with definitions by other parties, e.g., the manner in which different stakeholders or observers frame the radioactive waste management issues in Finland, or, how we frame the goals of coming together to discuss them in this forum.

As in other societal and study contexts, these concepts certainly had meaning here in the FSC Turku Workshop. They may form one set of tools with which we may consider further the background to a Decision in Principle, or stakeholder involvement. They may offer a lens through which the experience of the workshop itself may be perceived and evaluated. I found that one more concept (something to which linguists or anthropologists might refer as a *semantic polarity*) seemed to capture well the dynamic of our time together. That concept is: "inside-outside".

---

17. One aspect of my own demand was to gain insight on how *Finnish citizens' demand* upon the spent fuel management process has been interpreted.

Brought into the social psychological sphere, that polarity evokes our consciousness of "belongingness" (inside our group) and our perception of "otherness" (outside our group). That polarity may furthermore suggest issues of the *in-group* versus the *out-group*. In-groups and out-groups may be differentiated in terms of their features and those of their members, they may display: different social identities; perspectives; value systems; types of knowledge; roles; missions; resources; access to power; and relationship to other poles of power in society. The in-group enjoys a dominant position in defining the situation; the out-group is generally in a position of exclusion – although it may define its own power pole on that basis.

Such groups, and such distinctions, are familiar to all of us in our personal, professional, and social life. Each of us knows that it is possible to be a member of several groups at once, that these groups are more or less constituted or extensive, that the "in" or "out" status of any given group is in fluctuation, and that something of a struggle in maintaining or in changing status underlies other group activities. These realities, and the feelings and motivations they inspire – in the radioactive waste management field as in any other field of human endeavour – may be kept in mind as we look at social psychological issues arising in the workshop itself.

Indeed, I have made the choice in this thematic report to centre first on our own interactions, as a way of complementing the extensive analysis of Finnish stakeholder interactions developed here in Turku. By discussing the jokes and vocabulary heard in the workshop, I will address the difficulties of achieving an inclusive social dialogue on radioactive waste management. In a more direct response to the Finnish case study, I will suggest that greater stakeholder confidence in the very act of participation might be fostered by modifying consultation formats. Finally, I will offer an interpretation of how positions on radioactive waste came to change and evolve in Finland, and suggest that we should not curtail our reflection on the Finnish experience by attributing it to irreproducible cultural factors.

2.   Our jokes and their hypothesised function

The inside-outside polarity was called to my attention by the "jokes" offered by the different workshop speakers or moderators. (It seemed of value to be attentive to what made us laugh here, to the "glue" each person attempted to bring to our social well being).

Most of the jokes recorded below do not look fantastically funny on their own. In context, they did however make us laugh, and so they did each time allow us to express something of importance underlying our work, something of which each group member laughing had been dimly aware, something that sought the surface and was easily evacuated in that way.

Many of the jokes told during the workshop, I found, could be understood as touching upon a dynamic between inside and outside, or upon the difficulty of communicating among groups. Below they are classified in that light. If we assume, reasonably, that together we were seriously addressing our work, and that these jokes were not told for no reason at all, they may merit consideration, for what they can teach us of the core issues of stakeholder involvement.

2.1   *You have to stay inside, although you might want to escape!*

One of our Finnish hosts was pleased to notice the "excellent weather" – cold, wet, grey weather that would discourage the workshop participants from going outside during sessions. (Some murmured comments were heard: "the weather is not as bad as expected!").

A roundtable facilitator said his feedback to the workshop would be brief so that we could get out to the spa.[18]

A plenary speaker said he had tried, in preparing his presentation, to answer the questions posed in advance by the FSC, but noticed now that all the questions had already been answered by prior speakers; and so, we might all have gone swimming instead of listening to him.

These jokes seem to highlight an existential question of whether meaning, worth, and pleasure lie elsewhere: outside of the here and now, grappling with the issues of inter-stakeholder communication.

2.2   The difficulty of making the reality of one group apparent to another

A professor noted his impossible task: he had 20 minutes to explain how Finnish society works.

The language barrier was recognised: a plenary speaker found no word in Finnish for "stakeholder".

A representative of the safety authority was "shocked" by a ministry suggestion that nuclear power activity in the 1960s and 1970s had been "wild", unstructured, unsurveyed: "We were here", he protested. (This joke evokes issues of territory, recalling for instance that colonial powers often do not perceive the full presence of indigenous peoples.)

2.3   Suspecting that the status of one's own group is threatened – and reaffirming that status

A local elected official was "not so happy to be here" after learning that politicians hold the last place on a national survey list of trusted actors, headed by the police; however, as in his municipality "there are no police", he moved up a rank.

We had seen many large-scale maps allowing us to situate Finland in Europe, but upon which "Eurajoki is…nowhere"; to rectify this, he kindly showed a country map on which "we represent Eurajoki with a big dot" (indeed the only dot at all, if memory serves!).

2.4   Downgrading another group with a sharp characterisation

It was mentioned by Finns that the Swedes like to have an open process going on… but not to come to a decision (of course, this was not really a joke).

Downgrading one's own group characteristics (black humour)

"Maybe concepts like democracy, public participation, openness, transparency, and decision making seem complicated because… I am a political scientist."

---

18. A swimming pool/spa was a pleasant feature of the workshop hotel.

An inadvertent joke was achieved when someone revealed the feelings that might exist within the out-group: a burst of sympathetic laughter was heard when an opponent mentioned that seen from "below", the repository siting process had been "bitter".

2.5   *Lack of trust for knowledge or content far outside that of one's own group*

A parliamentarian ironically admitted that "the word I really loved in this process was 'bentonite'!".

2.6   *Someone made assumptions about other-group characteristics and got it wrong*

"When they told me during a site visit that bentonite is used in ladies' mascara, then I *really* knew I could not trust it!"

2.7   *No joke*

Perhaps a joke resided in the fact that Microsoft Word had automatically substituted, throughout a document distributed in the workshop, the word "scooping" for that of "scoping". (This pleads for caution in the blind application of procedures, even when they are designed to help…. Note that the word "siting" does not exist, either, in the Microsoft dictionary.)

2.8   *Failed expectations?*

A roundtable facilitator confessed, "We did not answer the discussion questions – we just spent the whole time talking to our stakeholder!" (as it happens, an eloquent opponent).

The humour of this joke resides in the idea that we may not have performed our task as required. However, its real power stems from the fact that of course, the entire workshop was indeed organised to allow this talking among parties. That is what we were supposed to do, what we were hoping to do. That is where the meaning, the worth and the pleasure lay: in talking with each other, and the more different that "other", the better. The "failure" to perform the task, paradoxically, and thus humorously, was the greatest sign of successful participation.

3.   The difficulty of reaching others

The jokes recalled above often revolved on the notion of how difficult it is to reach, or be reached by, others. Of course, there was no more difficulty in conducting exchanges at this workshop than in other social settings. Indeed, the participants were probably more successful in exchange than would be populations unaccustomed to international seminars with their multilingual and cross-cultural dimensions.

Still, the object of our workshop was to discuss stakeholder involvement, and the method chosen was to appeal to the widest possible range of stakeholder voices. On this criterion, a mixed result was achieved. Posiva, our organisers and hosts, deserve thanks for their work in structuring the programme, for a certain variety of voices was indeed heard. These speakers' thoughtful consideration of the process to date allowed their listeners to construct a many-faceted view of the Decision in

Principle and the Environmental Impact Assessment process. However, the list of Finnish invitees was dominated by institutional representatives (from ministries and agencies), and by professional observers (from universities and consultancies). A small number of opponent voices was heard, but one could ask: Where were the "private persons" mentioned, the representatives of the local press who seemingly had played such a strong role, other interested and affected parties?

The difficulty of identifying, contacting, and motivating the participation of stakeholders, of achieving representativeness, of finding ways to include others from outside the institutional sphere, should not be underestimated. One obstacle may lie, starkly, in conceiving of that otherness: that is the impression gained from examination of the list of international attendees. Almost no country represented in the Forum on Stakeholder Confidence used this opportunity to let its institutional actors travel and exchange, outside usual borders, with other stakeholders from their own setting. Despite the attractiveness of that concept, and the interest expressed for this opportunity at the January 2001 meeting of the FSC, finally the country representation at the Turku Workshop resembled a cautious first try. The Finns deserve all the more thanks for accepting, for us all, the risk of exposure.

The movement observed in society toward tolerance and seeking out of diversity, the movement in radioactive waste management toward inclusiveness, are at the source of the FSC initiative. This movement is still to be encouraged in the FSC work itself. On the basis of this first experiment, surely the FSC members can take courage indeed.

4.   Representing others: A matter of vocabulary?

The words we use express the manner in which we perceive the world; they place limits, as well, upon that perception. I wondered sometimes, in listening to workshop statements about stakeholder involvement, if we should not attempt radically new ways of speaking – and observe the consequences of these new framings for our radioactive waste management activities.

Many of the statements heard were extremely progressive if compared to attitudes and understandings heard in years or decades past from actors in the nuclear institution. They are the outcome of the sometimes painful confrontation between the technocracy and other spheres of society; many of the statements to which I reacted are indeed auto-reflexive, and make direct reference to this recent history of remoulding positions and understandings. If I highlight some statements, let it be seen as no sign of disrespect, but rather, as a benevolent attempt to reinforce the movement engaged (and sometimes difficult to maintain).

The vocabulary employed during the workshop to express the role of non-institutional stakeholders, despite the movement toward a new consciousness recognised above, sometimes seemed patronising. That means very literally that the institutional speakers, with no ill intent, expressed the hierarchical superiority of the institutional actor's position, point of view, or approach in regard to that of other actors.

Thus we heard from one roundtable that the EIA functioned not only as a framework, but also as a "guide" for active public participation. If this statement is taken to its logical conclusion, forms of participation and modes of expression that did not fit into the EIA might be considered only as "misguided". Such an implicit logic may subtly affect both the attitude of the institutional actors receiving input, and the willingness or ability of other stakeholders to provide input. Thus, while the roundtable statement welcomed the EIA, treating it as a "guide" might paradoxically contribute to restricting its effectiveness. In my view, EIA and social impact assessment are examples of tools responding to the challenge of representing impacts and views outside the delimited technological sphere. Instead of considering EIA to furnish now the enlightened guide, we may go on exploring even more ways of being guided in radioactive waste decisions by the "life outside EIA", as one university speaker put it.

"Decisions on technological management depend upon public support": this statement is lucid. But are we not now moving toward seeking public support in *formulating* those decisions, rather than seeking support for the implementation of decisions that are made?

"Authorities felt that Posiva's approach (in handling the EIA process) was adequate in creating trust among ordinary people": so much the better if this trust was created. Should the FSC, however, look primarily at ways of creating trust – thereby accepting a state of affairs in which the ordinary people are divorced from the decisions taken, as if their only mode of relation could be through trusting or not trusting?

It was reassuring for me to hear from one roundtable this observation: the Finnish Parliament's Decision in Principle signified that the deep disposition choice was in the good of society, and this level of decision thus implied that all of society must be considered in the unfolding of the management process.[19]

Today, the socio-technical object "radioactive waste" is construed and constructed in different manners – sometimes polarised – by different parties. Today's state of affairs is a *de facto* societal co-management of radioactive waste, in which the weight of the management burden is thrown upon the technologists' side. Redistributing this burden in society – so that it is carried in a mutually agreeable way – will require change in perception, and in vocabulary, on all sides.

5.   How to stimulate dialogue, participation, involvement?

No reader of these Proceedings will fail to perceive the relevance, and the difficulty, of "reaching others" through e.g., the Environmental Impact Assessment process. The question of how to stimulate dialogue, participation, and involvement among a broader range of stakeholders emerged as primordial in our roundtable discussions.

I was struck by contrasting responses given by different roundtables to this question addressed during the workshop session on EIA: "Stakeholders, did you get all the information you needed for your involvement?".

One response was: "Do not limit the information offered: we will choose what we take."

Another response was: "Too much information!".

This contrast in position alerted me to the role of information – here, mainly written information – in building bridges among different groups.

At various times during the workshop, stakeholders were described or even defined in terms of a knowledge criterion. For a ministry speaker, there "was not much stakeholder consideration

---

19. I would contrast this statement with one in the mid-1990s from a nuclear executive: since geological disposal had been found to suit a set of ethical objectives, he assured his audience at PIME (a European Nuclear Society public communications meeting) that they held "the moral high ground" in rejecting opposition to that option.

because there were no stakeholders" in the early years of Finnish nuclear-power generation, since "almost no one knew anything about nuclear power". (This speaker went on shortly to say that "the safety question caused almost no public concern": this juxtaposition suggests that the access to knowledge, and thus to concerned stakeholder status, by persons outside the nuclear power sphere would come only through accident or crisis – metaphoric or literal.) A deficit model appeared in STUK's characterisation of public knowledge of basic safety issues as "unfortunately very poor", and more or less "opposite" to that of natural scientists. Perhaps in a more neutral manner, a roundtable pointed out the difference between lay people and experts as being a difference of knowledge.

The public information offered to Finnish stakeholders (in particular, those outside the most directly active institutions) was meant to facilitate their involvement during the EIA process. It was an offer of the opportunity to access others' social knowledge, and to construct a new social knowledge inside one's own reference group.

It is hard to say, on the basis of what we heard in the workshop, whether a very broad range of stakeholders successfully exploited that opportunity. However, one issue of communication among knowledge bases stood out in the troubled observation voiced here: there was very little written input to the EIA process by members of the general public.

This "mystery", to which involved ministries and proponent institutions rightfully are attentive, suggests that the EIA should include modes of participation other than via the written word (of course, formulating a written comment was not the only possible way to manifest a point of view. Focused group interviews were conducted. There was a local EIA contact person in the municipalities under consideration for a spent fuel facility, and exchange with this person could probably resemble habitual modes of social communication.).

Looking deeper, the question posed by that "mystery" of low participation can be linked with our discussions of confidence-building measures. The general question is *whether there are modes of expression,* outside *written contributions, in which the different stakeholders could have confidence that they would affect repository decision making.*

My own first response to that question ignores perhaps the difficulties of faithfully incorporating unwritten, unsystematic knowledge into formal public decision. In reaction to the issue of low written participation, I wondered if there could not be something like a National Spent Fuel or Radioactive Waste day, a day whose free responses to the theme of waste, whose planned or unplanned community and individual "happenings", might be officially recorded by every journalist, every video amateur or professional, every poet or other interpreter who went into the streets or countryside. This proposal raises the same issues of participation, representativeness, integration and vulnerability to "hijacking" as does any other method of information collection in a policy-making goal. At least, though, it affirms that there may be many modes of relation to the issues underlying radioactive waste management.

Once again, observing our own behaviour here at the workshop may give us some insight on the difficulties of obtaining, or joining in, participation. We had three sessions of feedback from the roundtables. At each, a different system was used to call upon the next facilitator to step up to the podium. First we followed a rational order, the "natural" hierarchy of the numbers assigned to the tables. At the second session, a co-optation method emerged, with each facilitator at the end of his or her presentation choosing among the subgroups awaiting their turn. At the third session, a volunteer system was used, briefly. When there was hesitation among potential volunteers to come forward, we reverted to the more comfortable co-optation. This experience may remind us that it is sometimes very difficult to come forward, even when conditions are favourable, when we know we risk no rejection, and when we are sure that what we have to say is relevant. If we are convinced that we need to take into consideration a great diversity of public views in society's decision making upon radioactive waste, perhaps it may be necessary to ask again and again for volunteers to express their "critiques, questions and requests for additional clarification";[20] we may need the margin to wait for those volunteers to come forward.

One roundtable proposed this pragmatic position on participation in an EIA process: "We invite, and stakeholders choose to come or not". The verb "invite" suggests an interpretation: perhaps the party to which stakeholders are invited is no fun? In this connection, we may observe the very positive experience that seemed to be had in the Finnish process in regard to the role played by STUK, the safety authority. STUK were not "throwing the party"; they were simply – and actively – present, centred on creating a clientele who could be satisfied by receiving what they wanted, what STUK was perfectly poised to give, in terms of safety assurance. This experience suggests that it may be disruptive for the implementor to be charged with throwing the party of environmental impact assessment. This assessment has grown far beyond its original boundaries, and is asked now to contain an essential proportion of the total public consultation foreseen for radioactive waste management. Should implementors receive the backlash of any structural shortcomings in the consultation and decision process materialised by EIA? Should not other actors – in a legislative, governmental, or other mediating role – openly take responsibility for planning and throwing the party? The implementor's explicit role is one of proposing a technical design and operating concept that can stand up to criteria uncovered through the EIA; should not this role be independent of the conduct of that assessment? If the implementor – especially one intimately connected with the utilities generating nuclear waste – was relieved of conducting the EIA, would there be any positive impact upon the confidence of other interveners that their own participation would not be subtly discounted from the start?

6.   Spent fuel inside or outside

A Posiva director recognised the "lingering image question" that seemed to subsist in the communities considered for an underground facility. He had the sense that local concern for stigmatisation is an expression of some as-yet unformulated difficulty, reticence or discomfort with the idea of receiving spent fuel for storage. Indeed the "image" issue – here, the fear that identification with spent fuel will result in shunning of the community and in economic loss – translates all the uncertainty of how one's reference group is perceived by those outside, and of how to estimate, predict and control the impact of that perception upon their behaviour.

Can insight on this local apprehension be gained by looking at the Finnish national experience? Is spent fuel always seen as foreign to self-image? Is there a way to integrate spent fuel into the community without fear of stigma? I would state that incorporating spent fuel into the community space is a crucial issue, and that it is this issue that was present in some of the most significant radioactive waste management decisions taken in the past few years in Finland.

Upon the 1994 decision of the Finnish Parliament banning national import, and export, of spent fuel, stakeholders (national and local) found themselves with spent nuclear fuel resolutely inside their sphere. Formerly sent away (to the former Soviet Union), spent fuel now claimed status as part of their self-image. No longer could it be evacuated to an indeterminate outside space; a localised inner space had to be found for it. Stakeholders were confronted with two options: leaving the spent fuel

---

20. As a municipal council member from Eurajoki characterised typical local input to the EIA.

where it is (the so-called "zero-option" – which in this light does not seem likely to be a neglected option!), or, evacuating it to another inside – inside the earth.

This event seems to have shifted many balances. It produced, with some delay, the creation by the utilities of a dedicated organisation for the management of spent fuel: Posiva. More directly, it produced a swing vote in the Eurajoki municipal council. Three council members reconsidered their position on the prohibition of domestic spent fuel storage by the long-term municipal strategy document. The modified vote of two of these persons permitted a majority in favour of removing this clause.

It appears that the parliamentary debate on the import and export of spent fuel was introduced by the Greens, in response to a Norwegian minister's pointed question to his Finnish counterpart of whether evacuating spent fuel to Russia was really considered to be an acceptable solution. (One could say that the Finns saw their image as reflected by an outside stakeholder.) Studying the transcript of parliamentary debate, and any detailed record of Eurajoki municipal deliberations, might give insight into the reasoning and the process of attitude shift that lie, with other realities, at the root of the Finnish experience about which we are so curious.

7.   Finnish consensus culture?

Growing up in the United States, the meaning I learned to attach to the concept "consensus" was that everyone was supposed to form the same idea. A significant evolution of this concept was contained in the formula "agree to disagree".

When we were received by council members in Eurajoki before the workshop, many of us were impressed by the smooth co-operation among the representatives of different currents, despite their differing opinions on fundamental features of spent fuel management. In France, my home today, I believe that local council members of differing opinion would have seized such an opportunity to demonstrate with force their position, regardless of whether a prior vote, as in Eurajoki, had already settled the outcome of the debate. I mused that the Finnish representatives seemed to have "agreed to agree". This insight was rectified, at the end of the workshop, by the council member who had acted as translator for his colleagues. "Once the debate has taken place," he responded, "why bring it up again? We have a common objective, and so we are co-operating – going forward in parallel".

Often, as the roundtables evaluated what we were learning in the Turku Workshop, it was commented that the most intriguing features of co-operation among the Finnish actors simply had to be attributed to their "culture". As we felt we approached the core factors behind their decisions, suddenly our exploration was halted by the idea of "cultural differences". These loomed up as the basic explanation for why comparable events could not or would not take place in our own contexts.

But who can state, without long study and elusive measurement, the relative influence of "culture" and of other features of context and conjuncture – individual, organisational, political, social, historical – that determine decisions at any time? And did we not apply, sometimes too quickly, the lens of our own experience, our own implicit theories of culture, of what is possible and impossible, to our interpretation of what happened in Finland?

My own position is perhaps a little too "Swedish" for Finnish tastes. I believe that a continuing process of exchange among groups, a reciprocal attempt to grasp differing social identities, representations, roles, demands, and definitions of the situation, is vital to the iterative negotiation of a radioactive waste management system in which more and more partners may have confidence. In such a perspective, the visit of the FSC to Finland was a valuable initiative. To the extent that its lessons were limited by culture – inimitable Finnish culture or the very cultural identity of each visitor – it should not be the last such initiative.

THEMATIC REPORT ON COMMUNITY DEVELOPMENT AND SITING

A. Vári
Hungarian Academy of Sciences, Institute of Sociology

Summary

The paper analyses the Finnish spent fuel disposal facility siting from the perspective of community development, issues of fairness, and general factors of success. We found that anticipated positive impacts on host community development were the most important factors of local support. Second, the willingness of main stakeholders to adopt and combine several competing and changing concepts of fairness helped making legitimate decisions. Finally, we can conclude that in addition to important cultural factors which are unique in Finland, a number of siting elements have contributed to the success that are of cross-cultural nature.

The paper summarises the lessons learned about the Finnish spent fuel disposal facility siting process regarding the issues of community development, fairness, and the transferability of siting approaches across cultures. It is largely based on information presented within the framework of the OECD Forum of Stakeholder Confidence Workshop held in Turku, Finland, on 15-16 November 2001.[21]

1.   Community development

The planned spent fuel disposal facility is to be located near the Olkiluoto Nuclear Power Plant, on the Isle of Olkiluoto. The island belongs to the municipality of Eurajoki.

The territory of the municipality is 459 km^2, its population is roughly 5 900. Table 1 presents demographic data between 1950 and 2001. The data illustrate that the tendency of depopulation between 1950 and 1970 turned around in the 1970s, when the Olkiluoto Nuclear Power Plant was built and launched. However, the population started decreasing again in the late 1990s. The current rate of unemployment is around 14%.

In the 1970s agriculture was the dominant economic sector, employing about 37% of the active population. It was followed by industry and services in which 34% and 29% of the active population respectively were employed. By the 1990s, this tendency turned around, and recently only about 14% are employed in agriculture, 38% in industry, and 48% in services.

---

21. The author would like to thank the Finnish participants of the workshop for the valuable background data they presented in papers, round-table discussions, and personal conversations.

The Olkiluoto Nuclear Power Plant, owned by Teollisuuden Voima Oy (TVO) is the largest industrial facility in Eurajoki. It is currently the place of employment for approximately 600 people, roughly 160 of whom are the residents of Eurajoki. The appropriate management of spent nuclear fuel is crucial to the unhindered operation of the facility. Since Finnish laws prescribe that all nuclear waste generated in Finland should be disposed of in the country, the community of Eurajoki is strongly interested in finding an appropriate site for permanent disposal.

During the discussions about the possible siting of a disposal facility, it has become apparent that local supporters of the facility anticipate its positive effects on "employment and economy" (Lucander, 2001/a). In the current economic situation, there are not many projects of similar size and significance in Finland. According to some estimates, the planned repository would provide 100 new jobs on the average. It is hoped that the new facility will give a boost to the local economy.

Another significant factor for the positive attitude towards the repository is the familiarity with and a general confidence in the nuclear industry. Openness and excellent safety record of the nuclear power plant have largely contributed to this confidence. A further factor may be the increasing concern about the emission of greenhouse gases due to fossil fuel combustion. Some opinion leaders of the community have expressed that they considered nuclear energy as "green power" in comparison to fossil fuel energy production.

The majority of the local residents have confidence in the safety of the disposal technology to be applied. They appear to trust both the regulatory authority (STUK Radiation and Nuclear Safety Authority) and the company responsible for siting, constructing, and operating the facility (Posiva Oy). This, of course, does not mean that there has not been any concern regarding the impacts of the facility. Worries about the image of the region, the stigma effect, and adverse impacts on the market of local agricultural products were brought up by local opponents during the discussions. Another issue of concern was that of waste import. Some fear that EU legislation will be changed so that import and export of spent fuel will be permitted, and the new facility may accept foreign waste.

However, *the balance of anticipated positive and negative impacts appears to be positive.* This is shown by the results of a recent opinion survey, according to which 59% of the Eurajoki residents were willing to accept the repository (Lucander, 2001/b). Based on this support, the municipal council made a decision in favour of the project in January 2000. Presumably, considerations about the economic development of the community played a major role in this decision.

2.   Fairness of siting

Apparently, the majority of Eurajoki residents perceive the decision on siting the disposal facility in the community as fair. However, there have been individuals and movements in Eurajoki, in the neighbouring communities, and other candidate areas who have challenged the fairness of siting. In the following section, the debates on equity and fairness will be investigated.

2.1   Concepts of fairness

Competing concepts of fair distribution of social burdens and benefits lie at the heart of facility siting controversies. Linnerooth-Bayer and Fitzgerald (1996) describe three views of a fair approach to siting hazardous facilities: technical-hierarchical, individual-rights, and distributive justice views. The *technical-hierarchical approach* is characterised by strict central government pre-emption of local authority, limited public access, and a strong reliance on technical criteria. The *individual-*

rights approach shifts decision authority to the affected communities, and implies public participation and negotiation for compensation and incentives. In this case, local consent is the most important siting criterion. Supporters of the *distributive justice approach* strongly criticise the individual-rights approach because it inevitably leads to siting facilities near disadvantaged communities. Proponents of the distributive justice approach emphasise various fairness criteria, including, but not restricted to need, vulnerability, and responsibility (for the burden).

In addition to the siting approaches and the associated fairness criteria, there are also competing distributive principles which can be applied to site selection (Young, 1994). The principle of *parity* requires that all parties be treated in some sense equally. In the case of waste disposal, this may mean that all communities (counties, states, etc.) get equal shares of the burden. This can be implemented in various ways: e.g., the waste can be physically divided among communities; an equal-chance lottery can be organised to decide which community receives it; a rotational arrangement can be implemented where everyone is required to live near the facility some of the time; or compensation can be provided for the host community with other communities dividing the cost. A second distributive principle is *proportionality,* which means that the burden is distributed in proportion to certain fairness criteria (e.g., responsibility for the burden, endowment, etc.). This principle also can be implemented by the methods mentioned earlier, including – proportional – physical division, lotteries, rotation, compensation, etc. A third principle is *priority* where the burden, for example the waste, is allocated in whole to one community based on selected criteria.

There is a general agreement in the literature that *there is no single morally correct way for allocating scarce resources or burdens.* According to Hisschemoller and Midden (1989) what people consider "just" or "unjust" largely depends on the political system of which they are part. Linnerooth-Bayer and Fitzgerald (1996) suggest that competing perceptions of fairness are associated with plural world views defined primarily by social or group belonging. In addition, dominant views on fairness may vary over time within the same community.

2.2   *Changing concepts of fairness in the Finnish case history*

In the early 1970s, the siting of the Olkiluoto nuclear power plant was accepted by the municipality of Eurajoki under the condition that spent fuel would not be disposed of in the community. In 1978, when the operating license of the Olkiluoto power plant was issued, the intention was to send the spent fuel abroad for reprocessing, while the residual high-level waste was to come back for storage (Lucander 2001/a). However, the 1983 government decision obliged to prepare for final disposal as an option in addition to the alternative of reprocessing. The government also decided on milestones for the waste management programme, according to which a site for a final disposal facility should be selected by 2000 and operations should start by 2020 (Manninen, 2001). Although the 1983 decision did not exclude the export of spent fuel, in 1987 preliminary site investigations started and Eurajoki was among the five sites to be investigated.

According to Finnish law, generators are responsible for waste management and STUK is responsible for regulation and supervision of the nuclear waste facilities. In 1994 the *Nuclear Energy Act* was amended and as a result it banned the import of radioactive waste to Finland, as well as the export of waste after 1996. In 1995, TVO and Fortum Heat and Power (the owner of the second Finnish nuclear power plant, located in Loviisa) jointly established Posiva Oy to site, establish and operate a spent fuel disposal facility. Posiva selected the Olkiluoto site, and in 1999 the Ministry of Trade and Industry asked the municipality of Eurajoki to approve the siting. It also asked the government for a Decision in Principle that final disposal at this site "is in the overall interest of the society". The Decision in Principle was ratified by the Finnish Parliament in May 2001.

At the beginning of the "nuclear era" in Finland, sending the spent fuel abroad was seen as fair. It was assumed that for the Soviet Union economic benefits gained from the reprocessing were higher than social risks. This general view – reflecting the *individual-rights approach* – had changed by 1994, when a new amendment of the *Nuclear Energy Act* was passed which banned the export and import of nuclear waste. This change of legislation implies – at least on the international level – a *shift to the distributive-justice approach*, and the use of the responsibility criterion ('every country should handle its own waste'). The most important factors of this change include, but are not restricted to, the collapse of the Soviet Union and Finland's joining the European Union.

As mentioned earlier, it is feared that this approach may change again. During the siting debates some expressed their concerns about the future possibility of import:

- "Those with a negative attitude feared that ... foreign spent fuel might be imported to Finland" (Lucander, 2001/a, p. 2)

- "I have also a little suspicion that some day nuclear waste becomes commodity in the European Community and will be disposed to the remote, sparsely populated areas, perhaps in Finland?" (Tuikka, 2001, p. 2)

Since the export of spent nuclear fuel was stopped, it has been stored at the two nuclear power plants. This point-of-generation storage corresponds to the *distributive justice approach*, based on the responsibility criterion. On the other hand, site selection for the final disposal facility was dominated by the *individual-rights approach*. From the very beginning, a veto-right was assured for the municipalities, and a host community was chosen based primarily on the criterion of local consent.

However, the approach taken has been criticised for interpreting local consent too narrowly. Critics have questioned the fairness of the process because the consent of the *neighbouring communities* was not sought. For example, the Loviisa-movement rejected the narrow definition of the affected parties:

- "As the planned disposal affected not only Loviisa but also its surroundings the movement included representatives not only from the town of Loviisa but also from the four neighbouring municipalities (actually including even more distant ones). Also in this respect the movement rejected the definition made by Posiva." (Rosenberg, 2001, p. 1)

Although the siting process was dominated by the individual-rights approach, the situation is more complex. The host community happens to be the location of one of the two Finnish nuclear power plants, therefore *distributive justice, based on the responsibility criterion, also applies for Eurajoki* (but not for Loviisa). This has been emphasised during the debates by the supporters of the Eurajoki site:

- "Those with positive stand emphasised ... that Eurajoki having the benefits from nuclear power production has also the moral responsibility of wastes." (Lucander, 2001/a, p. 2)

At the same time, some opponents see this coincidence as unfavourable from the aspect of safety:

- "Those with a negative attitude feared ... too big concentration of nuclear activities." (Lucander, 2001/a, p. 2)

The distributive justice approach, based on the responsibility criterion, seems to be appealing for many. For example, in his presentation at the Turku Workshop, the leader of the opposition group in Loviisa claimed that his group "would rather store the spent fuel generated by the Loviisa power plant in their municipality, than move it to the planned Eurajoki repository". Similar ideas are expressed in his paper:

- "The movement .. succeeded in its main purpose, i.e., stopping the radioactive waste plans for Lovisa. But it did not succeed in the most important issue, i.e., resisting the whole idea of a final deposit. Turning it to the neighbour (in this case Eurajoki) was no victory in our mind." (Rosenberg, 2001, p. 2)

At the same time, some proponents argue that hosting a spent fuel disposal facility should not be seen as a burden, if compared to – the technically inferior – long-term storage. According to this argument, since the fuel has already been stored in Eurajoki and Loviisa, each of them would be better off with a safer permanent repository. It appears, however, that in the debates on long-term storage *versus* permanent disposal, fairness issues regarding *future generations* play a role that is of similar importance to that of technical safety.

Supporters of final disposal claim that this technology will help avoid imposing burdens on future generations. On the other hand, supporters of long-term storage argue that the latter technologies provide for a more equitable share of burdens, because they allow future generations changing the current management method and the allocation of waste (e.g., by implementing a rotational scheme). As a result of these debates, a compromise was reached, in terms of which permanent disposal would be applied, but the possibility of waste retrieval would be left open:

- "Upon termination of the final disposal phase of spent nuclear fuel, the repository can be permanently sealed… On the other hand, it will also be possible to retrieve the spent fuel from the repository to the surface." (Ministry of Trade and Industry, 2001, p. 3)

As we mentioned earlier, for most residents of Eurajoki the benefits of the facility (tax, jobs) offset the negative impacts. Besides renting a historical building by TVO, no other form of *compensation* has been offered to the host community. Some neighbouring communities, that anticipate more loss than benefit from the siting, raised the issue of compensation, but their claims were rejected. In general, compensation for unwanted facilities does not seem legitimate in the Finnish context.

The history of the siting process and the emerging debates reveal the plurality and changeable character of the principles and criteria of fairness. Due to their complexity, it is unlikely that a siting exercise will simultaneously meet all expectations. The study of ethics cannot be expected to provide formal rules that can be applied to derive a correct choice for the location of a facility (Massam, 1993). The way out of this dilemma is the exploration and open discussion of the views of main stakeholders regarding the principles and criteria of fairness, and the negotiation of an agreement. The Finnish siting process has been rather successful in this respect since the main stakeholders were willing to compromise on the most important issues (e.g., site location, management method). The combination of individual rights and distributive justice approaches proved to be acceptable for both the local community and the general public.

3.   Transferability of siting approaches

Political scientists emphasise the unique nature of the Finnish political culture as a major factor of successful[22] siting (Ruostetsaari, 2001). Undoubtedly, there are several characteristics of the Finnish context which were instrumental in this success, including general trust in technology and governments and the consensual style of decision making. However, I think that satisfying a number of cross-culturally valid provisions was of similar importance from the point of view of successful siting.

Based on an empirical research study investigating low-level radioactive waste disposal facility siting processes in five states of the United States, Canada, France, the Netherlands, Sweden and Switzerland, 10 recommendations were proposed for the design of siting processes (Vári et al., 1994). By analysing the Finnish case against the backdrop of the recommendations, we found that eight of them were met. These recommendations and the corresponding Finnish data are listed in the following[23]:

1.   The siting policy is integrated with broader policy regarding energy production and nuclear industry. A consensus needs to be established that the status quo is unacceptable; there is an important problem to be resolved; and the planned facility is the preferred solution to the given problem.

"In its report on energy policy approved in the autumn of 1997, Parliament emphasised the importance of the arrangement of nuclear waste management and the selection of a final disposal site for spent nuclear fuel in accordance with the valid schedule." (Ministry of Trade and Industry, 2001, p. 2)

2.   The goals of the waste management programme are clear. The source, type, and amount of waste to be disposed of at the facility should be well defined, and there need to be guarantees that no additional types and amounts of waste from additional sources will be shipped to the facility.

"On the basis of this Decision in Principle, final disposal facilities can be built for not more than the amount of spent nuclear fuel required by the final disposal needs assessed on the basis of the valid operating licenses of the existing Finnish nuclear power plants, such that the total amount of nuclear fuel to be disposed of will be about 4 000 tons at most." (Ministry of Trade and Industry, 2001, p. 8).

3.   Responsibilities for site selection, method selection, facility construction, operation, public education, and compensation are assigned to the same organisation. Prospects for success appear to be further improved if these responsibilities are assigned to a semi-private or private organisation and if the role of regulator is assigned to a separate public-sector organisation.

In order to carry out all tasks related to the long-term management of spent nuclear fuel, a jointly owned company, Posiva Oy was established by TVO and Fortum in 1995. STUK Radiation

---

22. In this paper successful siting is defined as a decision which is legitimate, as well as supported by host communities.

23. Recommendations cited from Vári et al. (1994) are typeset in italics. Data on the Finnish process are based on official documents and workshop papers.

and Nuclear Safety Authority is responsible for regulating and overseeing nuclear facilities (Ministry of Trade and Industry, 2001).

4.   Site and method selection do not occur simultaneously. The disposal method needs to be identified and made widely known before site selection.

"In accordance with Section 6 a of the Nuclear Energy Act, nuclear waste generated in connection with, or as a result of, the use of nuclear energy in Finland shall be handled, stored and disposed of in a way intended to be permanent in Finland." (Ministry of Trade and Industry, 2001, p. 2)

5.   The goal of the site-selection process is to identify a licensable site with host community support, rather than trying to identify the optimal site. A voluntary process in which communities are allowed to withdraw from consideration at any time further improves the chances for community support.

The Olkiluoto site was chosen from four candidate sites. Apparently, the candidate sites were rather similar in terms of technical safety, but host community support for the Olkiluoto site was the strongest. The right to veto was provided for the host communities during the site selection process.

6.   The goal of the method-selection process is to identify a licensable method with host community support, rather than trying to identify the optimal method.

"Along with other parties, the NGOs and other parties had an influence on the fact that Posiva did compare the alternatives more thoroughly in the assessment report and presented the reasons for eliminating alternatives. In addition, the NGOs also had an influence on the fact that retrievability and monitoring of the waste shall be studied in detail before granting the construction permit." (Leskinen, 2001, p. 2)

7.   The host community is directly involved in decision making regarding site selection, method selection, compensation and incentives. Local governments need to act as decision-making bodies, and local liaison groups need to facilitate public education and consultation.

"Routines for dialogue between municipality and utilities have been created to guarantee continuity. Thus several types of liaison groups were formed … So in its meeting on 24 January 2000 the Municipality Council accepted Posiva's Application." (Lucander, 2001/b, p. 1)

8.   The political leaders in both the legislative and executive branches of government display long-term commitment to siting a facility.

"While providing an absolute veto right to the proposed host municipality and requiring the final ratification by Parliament, the DiP (Decision in Principle) process has added a considerable amount of commitment to the process from the part of political decision makers, on both local and national level". (Vuori and Rasilainen, 2001, p. 2)

Table 1[24]

Population data of Eurajoki (1950-2001)

Year	Population	Year	Population
1950	6334	1997	6157
1960	5770	1998	6087
1970	5316	1999	6028
1980	5684	2000	5929
1990	6042	2001	5910

References

Hisschemöller, M. and Midden, C.J. (1989) "Technological Risk, Policy Theories and Public Perception in Connection with the Siting of Hazardous Facilities". In Vlek, Ch. and Cvetkovich, G. (eds.) *Social Decision Methodology for Technological Projects*. Kluwer Academic Publishers, Dordrecht.

Leskinen, A. (2001) "Scoping and Public Participation in EIA: Finnish Experiences". *The Process of Stepwise Decision Making in Finland: Past and Future of the Decision in Principle,* Second FSC Workshop, Turku, Finland, 15-16 November 2001. OECD/NEA, Paris, 2002.

Linnerooth-Bayer, J. and Fitzgerald, K.B. (1996) "Conflicting Views on Fair Siting Processes: Evidence from Austria and the US". *Risk Health, Safety and Environment*, Vol. 7, No. 2

Lucander, A. (2001/a) "The Process of Stepwise Decision Making in Finland: Past and Future of the Decision in Principle". *The Process of Stepwise Decision Making in Finland. Ibid.*

Lucander, A. (2001/b) "Confidence Building: What Gives Confidence to the Various Categories of Stakeholders?" *The Process of Stepwise Decision Making in Finland. Ibid.*

Manninen, J. (2001) "Historical and Legislative Framework of the Decision in Principle". *The Process of Stepwise Decision Mmaking in Finland. Ibid.*

Massam, B.H. (1993) *The Right Place*. Longman, Singapore Publishers, Singapore

Ministry of Trade and Industry (2001) "The Decision in Principle by the Government Concerning Posiva Oy's Application for the Construction of a Final Disposal Facility for Spent Nuclear Fuel Produced in Finland".

Rosenberg, T. (2001) "What Could Have Been Done? Reflections on the Radwaste-battle, as Seen from the Bottom". *The Process of Stepwise Decision Making in Finland. Ibid.*

Ruostetsaari, I. (2001) "Explaining the Ratification of Nuclear Waste Disposal by the Finnish Parliament: Political Culture and Contextual Factors". *The Process of Stepwise Decision Making in Finland. Ibid.*

---

24. Source: Sahkoisen elinvoiman, Eurajoki, 2001.

Tuikka, K. (2001) "The Opposition's Turn to Speak". *The Process of Stepwise Decision Making in Finland. Ibid.*

Vári, A., Reagan-Cirincione, P., and Mumpower, J.L. (1994) *LLRW Disposal Facility Siting: Successes and Failures in Six Countries.* Kluwer Academic Publishers, Dordrecht.

Vuori, S. and Rasilainen, K. (2001) "The Role of the Public Sector's Research Programme in Support of the Authorities and in Building Confidence on the Safety of Spent Fuel Disposal". *The Process of Stepwise Decision Making in Finland. Ibid.*

Young, P. H. (1994) *Equity in Theory and Practice.*

THEMATIC REPORT ON STRATEGIC DECISION MAKING

T. Isaacs
Lawrence Livermore National Laboratories, USA

This Second FSC workshop focused on, "Understanding the factors that influence public perception and confidence in the area of radioactive waste management…" The workshop was held in Finland in close co-operation with Finnish stakeholders. This was most appropriate because of the recent successes in achieving positive decisions at the municipal, governmental, and Parliamentary levels, allowing the Finnish high-level radioactive waste repository programme to proceed, including the identification of a proposed site. The workshop objective was to gain insight in answering the question, "How did this political and societal decision come about?"

1.   The importance of public involvement in building public confidence

It is clear that while governments and private organisations have responsibilities in carrying our radioactive waste programmes, societal consent is still required. Thus there remains an important role for the public, and an open and transparent process is necessary to maintain their support. Often the public will support decisions that they may not agree with if they believe the process in reaching the decision was fair.

Much of the workshop discussion appropriately focused on the roles of public participation and public communications in building public confidence. It was clear that well constructed and implemented programmes of public involvement and communication and a sense of fairness were essential in building the extent of public confidence needed to allow the repository programme in Finland to proceed.

In particular, consistent with Finnish culture and institutional and political arrangements, there appears to be a focus on local and to a lesser extent, regional involvement. Numerous Finnish examples of public participation elements implemented over significant time periods demonstrated the value of such programmes. Indeed they continue to be essential for public acceptance to be achieved and then sustained. It was clear that the emphasis is on ensuring an appropriate and fair process, not just on developing a project.

Significantly enhancing this has been the development, public involvement, and subsequent approval of the Decision in Principle, providing the clear national agreement on the need for disposal. This, accompanied by clear roles and responsibilities among the implementor (Posiva), regulator (STUK), the public, municipalities, government and parliament, have defined a process and a rationale that have led to increasing confidence over time.

It was also clear that there were a number of other elements beyond public involvement that contributed substantially to the success in Finland to date. And, in fact, it appeared that these other factors were also necessary to achieving the Finnish public acceptance. In other words, successful public participation and communication were necessary but not sufficient. What else was important?

By agreement, the other rapporteurs focused their summaries on issues surrounding public participation and issues of fairness. So my remarks focus on organising these other factors that we had heard so compelling described by the Finnish stakeholders into a set of "lessons learned" that might have broader application.

2.   What else creates public confidence?

In addition to well-planned and executed programmes of public participation and communication, there appeared to be three major additional elements that significantly contribute to building and maintenance of public confidence. These may apply both in Finland in general and in the municipality of Eurajoki, which has volunteered to host site characterisations to determine if a repository can be suitably built there.

Summarised, they are:

competence: The implementors and regulators are seen as competent and have demonstrated competence over an extended period of time;

good intentions: The implementors, regulators, and other major participants are seen as well intentioned and wanting to do what is in the best interests of the host municipality in particular, and the general population;

a willingness to change to meet public concerns: Implementors, in particular, are willing to engage affected communities in frank and open discussions. They are interested in understanding the concerns that might exist and are willing and flexible enough to change programme elements to deal with such concerns.

3.   Competence

We learned that Finnish culture is most often based upon consensual decision making. They demonstrate an impressive ability to discuss contentious issues fully, and to disagree, but then reach political consensus. And once they do, the culture is such that all then take part is seeing that the decision is implemented as effectively as possible.

Municipalities play a central role and must say yes if siting is to occur. Beyond that, the "State" has dominance in many matters and the political elite tend to know each other and in some senses operate as a "club."

Importantly in this framework, there appears to be a high level of trust in institutions such as the police armed forces, and church. And there is an inherent confidence that science and technology, put to appropriate uses, can help solve most problems. Those responsible in Finland for nuclear activities seem to enjoy much of the same confidence.

The Finnish nuclear experience supports this confidence. They have a fine track record in the application of science and technology and exhibit a national pride in Finnish technical capabilities. Specifically, they have had positive experiences to date in the operation of the four Finnish nuclear power plants and the low and intermediate level waste facility that is in operation. It is no accident that two of the reactors and the waste facility are located in the volunteer municipality of Eurajoki. They are familiar with nuclear projects, citizens in the community work at these facilities, and they have confidence that the implementor (Posiva) and regulator (STUK) know what they are doing, and will do what's necessary to assure safety and protect the citizenry.

Posiva and the other nuclear organisations have been most willing to develop these capabilities and demonstrate competence with step-wise decision making. Both in the sequential development of the nuclear facilities in Eurajoki municipality and in the larger National nuclear and repository programmes, they are willing to take numerous sequential steps and the time necessary to earn the public confidence that comes with doing each successive job well.

4.   Good intentions

Competence alone does not guarantee confidence. The public must also believe that the involved parties have the citizenry's best interests foremost in mind as they move forward.

Here, too, the Finnish culture and experience provides a foundation for public confidence. There is an emphasis on safety in society, both for local affected communities and for the common national benefit.

Perhaps most important is the absolute veto of the potential host municipality. By guaranteeing the siting of a repository only where it is wanted, the process builds in a very high degree of control by those most affected and assures that their interests and concerns will be carefully addressed.

Decisions have been made with such priority in mind both at the national and municipal level. In 1984 the strategic decision was made with the objective of finding a suitable site by 2000. In 1994, importantly, there was a decision to stop exporting spent nuclear fuel and to accept responsibility for ultimate disposal within Finland. These commitments have been taken seriously, hence the recent decisions.

There was also a 1993 Parliamentary decision to reject a new nuclear power plant, demonstrating again that new facilities are not inevitable, but should be considered against the Finnish needs. While the authorities are, at present, once again considering the need for an additional plant, the decision on the future of nuclear power is not linked directly to the current waste decisions, helping to keep them from being overly politicised.

Thus numerous stepwise developments all provided the confidence to the public that reasonable next steps in approving a Decision in Principle and a municipality acceptance for repository development would keep public safety and overall societal good pre-eminent. Among these were:

- The sequential development of four successful nuclear power plants.

- The construction of the existing low and intermediate level waste repositories.

- The existence of facilities for spent nuclear fuel storage.

- The decisions to neither export nor import spent fuel.

- The rejection of a proposed new nuclear plant.

- The absolute veto authority of any municipality in siting the repository.

Adding to this sense of confidence is the commitment for many future steps and decisions before final decision is taken to construct, operate, and ultimately close the repository. Integral to this progress is the explicit option for waste retrievability in the future. And the municipality also relies on its confidence in the regulator to oversee the developments and to have their best interests at heart. In fact this confidence is so strong that there is little evidence of public concern about long term safety of the repository; they are confident that those in charge will either assure long term safety or they will not build the repository.

5.   Frank discussions and a willingness to change

The third element that appeared to be important in building the public confidence was a commitment to continuing meaningful discussions that were truly two-way. The major organisations responsible for implementation and oversight wanted to not only inform the public of their decision and plans, but to engage them such that public concerns and interests were identified and dealt with in a proactive manner.

Given the autonomy of the municipality in Finland, there is a priority with the implementor and regulator to "satisfy your customers needs. And in this case, the customers are principally the local public and local decision makers." The regulator in particular, is seen as on the side of the municipalities and reliable. To demonstrate their commitment, the highest level of STUK management is quite visible in the municipality and works hard to ensure that the process fully engages the local public and its elected representatives. They and the other participants realise that building and maintaining trust takes time and requires the successful completion of many steps and continual dialogue. They seem committed to do what it takes.

There were many chances to erode public confidence. In particular, the initial municipal decision did not accept the concept of spent fuel disposal, a decision that was subsequently reversed over a period of years. But the stepwise, transparent, and open process, particularly with the affected municipalities and their citizens has kept progress on track. This is aided in no small degree by the mature and thoughtful approach taken by virtually everyone we met from Finland. As was said, debate takes the form of "enlightenment by intellectuals," not by a political and media circus. All of the Finnish individuals with whom we met, including those sceptical of or against the repository programme, were able to artfully express their views and have them taken seriously and with respect.

Since there is a general sense that the regulator and implementor are competent and well intentioned regarding assuring the performance of the repository for geological time periods, the local community appears to be most interested in "above ground, every day things." And when it comes to impacts, "citizens are the experts of local questions." The emphasis is less on allaying concerns than on fixing them.

Therefore, the focus has been on safety and municipal needs. Interestingly, there is no provision for compensation to a host community, something that is expected as a part of many other national programmes. Yet Posiva has worked carefully with the municipality of Eurajoki to develop a win/win arrangement. A current home for the elderly, housed in a historic building, will be renovated and then used to house programme officials while the rent is devoted to constructing a new, modern facility for the elderly. This type of careful and thoughtful co-operation appears to build not only a sense of fairness, but of shared ownership.

6. Conclusion

Culture, politics, and history vary from country to country, providing differing contexts for establishing and maintaining public confidence. What works in one country will not necessarily be effective in another. Nonetheless, there appear to be certain elements that might be common to programmes that are successful in sustaining public confidence. These elements were clearly on display in Finland.

The need for the programme is clearly established:

- Roles and responsibilities of the players are well understood.

- Respect of the need for societal consent is apparent.

- A clear, open, and transparent process is used in decision making.

There are many sequential steps taken as the programme unfolds that include the possibility of altering or reversing course.

Programme officials recognise that due deliberative process takes time and are willing to invest the time.

In addition, these three factors also may be important to achieving public confidence and support even when the above factors are evident:

- Responsible organisations are seen as competent by the public and have demonstrated their competence.

- Responsible organisations are believed to be well intentioned, that is to have the best interests of the public at heart as they implement their programmes.

- Responsible organisations are willing to engage in frequent frank discussions with stakeholders and to adapt programme decisions to deal directly with stakeholder concerns and considerations.

PROGRAMME OF THE WORKSHOP

Inauguration of the Workshop

Presentation by **C. Kessler**, Deputy Director-General of NEA.

Session I

Background to the Decision in Principle

Chairperson: **C. Létourneau**, Department of Natural Resources, Canada

Explaining the Ratification of Nuclear Waste Disposal by the Finnish Government
I. Ruostetsaari, Senior Researcher (Academy of Finland) and Docent, University of Tampere, Finland.

Historical and Legislative Framework of the Decision in Principle
J. Manninen, Deputy Director-General (Nuclear Energy), Ministry of Trade and Industry (Energy Department), Finland.

Safety: One Crucial Element for Decision Making
J. Laaksonen, General Director, STUK (Radiation and Nuclear Safety Authority of Finland).

Session II

The Process of Stepwise Decision Making in Finland:
Past and Future of the Decision in Principle

Chairperson: **J. Lang-Lenton**, ENRESA, Spain
Moderator: **M. Aebersold**, Bundesamt für Energie (BfE), Switzerland

Implementor and stakeholder discussions on:

- What triggered your involvement? What were your concerns, aims, and hopes?

- What were the main steps, when were you involved (site selection, Environmental Impact Assessment EIA, etc.)? With hindsight, should you have become involved at a different time?

- What were the criteria you used to move forward?

- Who were the relevant contacts (local or national government, implementors, others) in your decision-making process?

- What are the results of your involvement?

- What comes next? What undesirable events do you fear, if any?

Stakeholders voices

Past and Future of the Decision in Principle: The Implementor's Point of View
V. Ryhänen, Managing Director Posiva Oy, Finland.

The Regulator's Relationship with the Public and the Media: Satisfying Expectations
T. Varjoranta, Director, Nuclear Waste and Materials Regulation, STUK, Finland.

The Opposition's Turn to Speak
K. Tuikka, Representative of "Kivetty liike" (opponent group to the investigation site), Finland.

The Process of Stepwise Decision Making in Finland: Its History and Outcome
A. Lucander, Member of the Eurajoki Municipality Council, Finland.

Roundtable discussions of information presented in the foregoing presentations, focusing on the following questions:

- What were the most important steps in the decision-making process for the different stakeholders?

- What influenced the process and the outcome (Decision in Principle mechanism, Posiva's work, your involvement, something else)?

- What are the lessons learnt?

Session III

Stakeholder Involvement, particularly in the Environmental Impact Assessment (EIA)

Chairperson: **P. Ormai**, Public Agency for Radioactive Waste Management (RHK), Hungary
Moderator: **H. Sakuma**, Japan Nuclear Cycle Development Institute

Stakeholder discussions on:

- Who were the most important stakeholders?

- What information was provided, and by whom?

- What was the role of the EIA for your involvement?

- What impacts was assessed to the environment, the economy, the regional image, etc. and how have they been accepted?

- What were other opportunities for your involvement?

Stakeholder voices:

Stakeholder Involvement, Particularly in the Environmental Impact Assessment
A. Väätäinen, Senior Adviser, Ministry of Trade and Industry (Energy Department), Finland.

Public Participation in the Environmental Impact Assessment: One Alternative of Involvement
P. Hokkanen, Researcher, Publicly Administered Nuclear Waste Management Research Programme, Finland.

Stakeholder Involvement, Particularly in the Environmental Impact Assessment
J. Jantunen, Senior Adviser, Finnish Environment Administration.

Stakeholder Involvement in Posiva's Environmental Impact Assessment
J. Vira, Research Manager Posiva Oy, Finland.

What Could Have Been Done? Reflections on the Radioactive Waste Battle as Seen from Below
T. Rosenberg, Representative from a site investigation municipality (Loviisa).

Scoping and Public Participation in the Environmental Impact Assessment
A. Leskinen, Researcher on EIA, Diskurssi Oy, Finland.

Roundtable discussions on the following questions:

- Was the stakeholder involvement process sufficient?

- Did you get all the information you needed for your involvement?

- What are the lessons learnt?

- How could your involvement be improved in the future?

- Survey/feedback of roundtable discussions.

Session IV

Confidence Building: What Gives Confidence to the Various Categories of Stakeholders?

Chairperson: **D. Appel**, PanGEO, Germany
Moderator: **S. Webster**, European Commission

Stakeholder discussions on:

- What measures were taken and by whom to build confidence of the different stakeholders?

- How effective were these measures and how did they influence trust?

- What contributed to increase confidence especially among the public?

- How did perceptions and expectations develop?

- What is more important: technical analyses or the value of the actors?

- What do the specific stakeholders require from the other stakeholder over the long term (government, Posiva, etc.)? What organisations are important for building trust?

Stakeholder voices:

What Gives Confidence to the Various Categories of Stakeholders?
J. Andersson, M.P., Member of the Green Parliamentary Group, Finland.

Confidence Building: What Gives Confidence to the Various Categories of Stakeholders?
A. Lucander, Member of the Eurajoki Municipality Council, Finland.

Social Science and Nuclear Waste Management
T. Litmanen, Senior Assistant (Researcher on communications), University of Jyväskylä, Finland.

The Role of the Public Sector's Research Programme in Support of the Authorities and in Building Confidence on the Safety of Spent Fuel Disposal
S. Vuori, Research Manager, VTT Energy, and **K. Rasilainen**, VTT Energy, Finland.

Roundtable discussions to consider the following questions:

- What was (or would be) important for developing your confidence: technical aspects (scientific expertise, long-term safety retrievability, control, etc.); social aspects (ethics, openness, financing, etc.); institutional aspects (institutional framework); other?

- How would you rank the various confidence-building measures you have heard about?

- What were positive and negative experiences for gaining confidence and trust?

- What are the lessons learnt? What should be done to improve confidence and trust?

- Survey/feedback of roundtable discussions.

Session V

Conclusions, Assessment, and Feedback

Chairperson: **Y. Le Bars**, President of Andra, and
Chairman of the Forum on Stakeholder Confidence (FSC)

Reports from thematic rapporteurs on what they have observed and learned during the Workshop.

I. Public Governance
F. Bouder, OECD/PUMA.

II. Social Psychology
C. Mays, Institut Symlog, France.

III. Siting and Community Development
A. Vári, Hungary Academy of Sciences.

IV. Strategic Decision Making
T. Isaacs, Lawrence Livermore National Laboratories, USA.

Feedback from the participants in the Workshop.

Closing remarks.

BIOGRAPHICAL NOTES OF WORKSHOP SPEAKERS AND ORGANISERS

Michael AEBERSOLD

M. Aebersold is a scientific expert official within the Swiss Federal Office of Energy. His main areas of responsibility include nuclear waste disposal policy, decommissioning and waste disposal fund, and nuclear fuel cycle and non-proliferation. He was Switzerland's representative for the preparation of the Joint Convention on the Safety of Spent Fuel Management and on the Safety of Radioactive Waste Management. He has been and still is involved with the Federal Working Group on Nuclear Waste Disposal (AGNEB), the Wellenberg Technical Group, the Wellenberg Economic Working Group, the Energy Dialogue Working Group and the Expert Group on Disposal Concepts for Radioactive Waste (EKRA). His technical background is in inorganic and physical chemistry.

Janina ANDERSSON

Ms. Andersson has been a member of the Finnish Parliament since 1995 and is currently serving as President of the Green Parliamentary Group (running from 2001-2003). She is a Master in Political Science (M.Pol.Sc.) and has special interests in environmental policy; the Baltic Sea, energy policy.

Detlef APPEL

Free-lance geologist working mainly in the fields of groundwater protection and final disposal of radioactive and "conventional" waste. More specifically in the radioactive waste area, he is: (i) consultant to communities, political parties, non-governmental organisations and authorities (German federal government and *Länder* governments in licensing procedures or in hydrogeological) questions related to final disposal; and (ii) member of the German Commission on Reactor Safety (Board on Supply and Disposal), the BMU Working Group on Methodology of Disposal Site Selection, and the Swiss Expert Group on Disposal Concepts for Radioactive Waste. Special interests relate to the methodology of site selection and assessment.

Frédéric BOUDER

Frédéric Bouder has been an administrator in the Public Management Service (PUMA) of the OECD since 1996. He is working on a range of issues concerning most aspects of public sector innovation as well as policy challenges and governance trends. Since 1998, he has been responsible for policy coherence and horizontal projects in the division on governance and the role of the state. His main fields of activity cover Governance for Sustainable Development and Risk Management in the

Public Sector. Before joining the OECD, Frédéric Bouder worked as a researcher and consultant in Public Management. Frédéric Bouder was educated at the Institut d'Études Politiques of Paris and at the University of Paris I Panthéon-Sorbonne, where he completed postgraduate research in comparative law.

Pekka HOKKANEN

M.Soc.Sc. Pekka Hokkanen is working as a project manager in the Department of Political Science and International Relations in University of Tampere. He has been taken part in the Publicly Administrated Nuclear Waste Research Programme (JYT2001) since 1997. His studies deals with the EIA-process and especially with the public participation. His ongoing doctoral thesis concentrates on the relationship between EIA and decision making.

Thomas ISAACS

Tom Isaacs is the Director of Policy, Planning, and Special Studies of the Lawrence Livermore National Laboratory, in California. Much of his career was spent in the US Department of Energy and the Atomic Energy Commission working on matters related to the development and application of nuclear power, waste management and disposal, and public acceptance. He was the chairman of the NEA Expert Group which produced "Nuclear Education and Training: Cause for Concern?" and was the lead US delegate to the NEA RWMC for a number of years. He is a member of the US National Academy committee on the development of nuclear waste repositories.

Jorma JANTUNEN

Between 1983 and 1995 held different positions in local and regional environmental administration, mostly connected with nature conservation and with environmental impact assessment. Since 1995 has been Senior Adviser (EIA-co-ordinator) for the Uusimaa Regional Environment Centre and Senior Adviser (EIA-co-ordinator) for the Finnish Ministry of the Environment, Land Use Department from 1999-2000. Has a Master in Nature Sciences from the University of Helsinki.

Carol KESSLER

Carol Kessler became the Deputy Director General of the Nuclear Energy Agency in August 2001. Prior to that she was employed by the US Department of State for 15 years, most recently as the Senior Co-ordinator for Nuclear Safety. From 1994-2000 she represented the US government in the G-7 Nuclear Safety Working Group. Her area of specialisation is Soviet-designed reactor safety.

Jukka LAAKSONEN

Prof. Jukka Laaksonen has worked since 1997 as Director General of STUK, which is the Finnish regulatory organisation for radiation and nuclear safety.

Prof Laaksonen started his career as researcher in reactor physics area with VTT, Finnish Technical Research Centre, but has held various positions with STUK already since 1974. In 1981-82 he worked with the reactor systems branch of US Nuclear Regulatory Commission, and in 1987-89 he worked as senior officer with the IAEA.

Prof. Laaksonen has chaired a number of international meetings, working groups and committees in the nuclear safety field, and is currently Chairman of the NEA Committee for Nuclear Regulatory Activities, as well as Chairman of the EC Regulatory Assistance Management Group that supports regulatory organisations in the Central and Eastern Europe.

Jorge LANG-LENTON

Mr. Lang-Lenton studied Industrial Engineering followed by a Masters in Nuclear Engineering from the University of Madrid. He went on to receive a Master in Business Administration from the University of Manchester.

He worked for 9 years for ENUSA, Nuclear Fuel Manufacturing. He joined ENRESA (Radioactive Waste Management) in 1985, and currently is the Communication Director of the company.

Yves LE BARS

Yves Le Bars is President of Andra, the French National Radioactive Waste Management Agency. Before joining Andra in January 1999, he was Director-General of BRGM, the French Geological Survey Service. His background is in Water and Forestry Engineering. He specialised in this discipline at the École polytechnique, became Head of Urban Planning Services of the city of Grenoble and was, for 11 years, Director-General of CEMAGREF, the French national body responsible for environmental and agricultural engineering research. He has also held the position of Technical Counsellor at the French Ministry of Agriculture. Mr. Le Bars is currently the Chairman of the NEA Forum on Stakeholder Confidence.

Antti LESKINEN

Studies on EIA since 1986. Researcher and teacher of planning and decision making at the University of Helsinki. Academic dissertation on "Environmental Planning as Learning" 1994, which combines planning and organisational theories on environmental management .

Holds the first university post of EIA in Finland ("docent"[1]) since 2001 at the University of Tampere, while working mainly at Diskurssi Ltd . Consultant in the European RISCOM II - programme on public participation.

Carmel LÉTOURNEAU

After completing her Master degree in Environmental Sciences, Ms. Létourneau embarked on a 20-year career in the field of nuclear energy, first as a health physicist with the federal nuclear

---

1. A docent in a unique Scandinavian university post, with the right to teach and e.g. examine academic dissertation. However a docent works often outside the university and does not take part in the university administration like a professor.

regulator, then as a Senior Policy Advisor with the federal department of Natural Resources in the Energy Sector. She heads the team proposing Government of Canada policy and legislation on the long-term management of nuclear fuel waste.

Tapio LITMANEN

Dr. Tapio Litmanen works as a senior lecturer in Sociology at the University of Jyväskylä. His special interest areas are environmental sociology and sociological risk research. He took a part to the multiphase research programme, which aim was to support Finnish authorities in their activities concerning spent fuel management. In his doctoral thesis Dr. Litmanen studied nuclear technology from the point of view of sociology.

Altti LUCANDER

Studies: 1960 MSc HUT techn. fys.

Mr. Lucander has been a Member of Eurajoki Municipality Council since 1989.

Career: Oy Control Ab, I&C sales and design 1961-67. Rauma Repola Oy Pori Works, I&C chief design engineer 1967-73. TVO I&C manager 1973-94, training manager 1994-2000. Independent consultant 2000-2001, retired 2001.

Jussi MANNINEN

Mr. Manninen graduated in 1965 in reactor physics from the Helsinki University of Technology, where he also took a degree to be a Licentiate of Technology.

Since 1966 he has held different positions in the Helsinki University of Technology, Ministry of Trade and Industry and Ministry for Foreign Affairs, all connected with energy questions and mostly with nuclear energy.

Since 1993 he has been Deputy Director General (nuclear energy) in the Energy Department of the Finnish Ministry of Trade and Industry.

Claire MAYS

Claire Mays is a social psychologist (Institut Symlog, Cachan, France). She has conducted research and intervention in nuclear power plant worker safety, risk communication in the radioactive waste field, and social and psychological impacts of large radiological accidents. Claire is a member of a US National Academy committee on technical and policy challenges of geological disposition of high level waste.

Peter ORMAI

Peter ORMAI has a degree in chemical engineering and a PhD in the field of radiochemistry. His professional career in nuclear industry began in 1977 when he joined the Hungarian nuclear power plant at Paks. In 1982 he became head of radiation protection laboratories and in 1994 was appointed as head of Radioactive Waste Management Department. Prior to taking up his current post as a chief scientific engineer at the newly formed Radioactive Waste Management Agency (PURAM) he acted as a senior expert in the Hungarian radwaste disposal projects. He has been representative of several international projects and acting as invited expert in working groups (IAEA, EU, OECD/NEA).

Claudio PESCATORE

Claudio Pescatore is Principal Administrator for radioactive waste management at the OECD/NEA. Active at national (USA, Italy) and international level (OECD/NEA) in technical, strategy, and policy areas dealing with radioactive waste, he has over 20 years' experience in the field, and a multifaceted career as programme manager, consultant to industry, consultant to R&D agencies and safety authorities, university lecturer, and researcher. His background is in the physical sciences with a doctorate in nuclear engineering.

Kari RASILAINEN

Kari Rasilainen is specialised in the performance assessment of nuclear waste disposal (Technical Research Centre of Finland). He is engineering physicist by education and his doctoral thesis (1997) discussed the testing of performance assessment models with the help of natural analogues. Currently he is the co-ordinator of a new four-year (2002-2006) research programme in Finland called the National Research Programme on Nuclear Waste Management (KYT).

Thomas ROSENBERG

Thomas Rosenberg (born 1953) is a sociologist (M. Pol.Sc.) formerly working at Åbo Akademi, the Swedish university i Turku/Åbo, mainly interested in sociology of science and language issues. Now editing Nordic Studies on Alcohol and Drugs, a scientific journal published by Stakes, the National Research and Development Centre for Welfare and Health. At present off duty, working with a 3 year project concerning the future prospects of the Swedish minority in Finland. Since the 1970s engaged in environmental issues, and during the late 1990s chairman of the citizen movement against disposal of radwaste in Loviisa. Regularly writing columns and articles in different newspapers and magazines, mostly in Swedish.

Ilkka RUOSTETSAARI

Education: doctor of social sciences (political science) 1989. Doctoral thesis concerned determination of Finnish energy policy. He has also studied deregulation of Finnish electricity markets (1998).

Ruostetsaari has held various academic posts, e.g., professor of political science, since 1985 at Department of Political Science and International Relations, University of Tampere. Now he is working as a senior research fellow at Academy of Finland.

He is chairman of a co-operation group appointed by Ministry of Trade and Industry for social science studies on nuclear waste disposal.

Veijo RYHÄNEN

Mr. Veijo Ryhänen is Managing Director of Posiva.

He graduated from the Helsinki University of Technology in 1974 and holds the degree of Master of Science (Eng.) in Technical Physics.

He commenced his professional career in the field of nuclear waste management as a Research Scientist at the Technical Research Centre of Finland in 1975. Mr. Ryhänen joined the utility TVO in 1977, and he was involved in different project tasks related to spent fuel management, final disposal of operating waste and decommissioning of nuclear power plants. In 1986, he was appointed Manager for Nuclear Waste Management.

As TVO and FORTUM, the owners of the Olkiluoto and Loviisa nuclear power plants, established a joint waste management company Posiva, Mr. Ryhänen was appointed the company's Managing Director in 1995.

Hideki SAKUMA

Hideki Sakuma is Senior Research Co-ordinator at the Japan Nuclear Cycle Development Institute (JNC) His current responsibility includes co-ordination of JNC's national and international research, and development of projects in both technical and non-technical areas. His involvement at t the international level started in the mid-1980s with membership in the Joint Technical Committee of the OECD/NEA International Stripa Project. His other experiences include membership in the OECD/NEA International Review of SR97, a safety study prepared by SKB for a geologic repository of spent fuel in Sweden. His technical education background is Oceanography.

Kimmo TUIKKA

Mr. Kimmo Tuikka took his matriculation examination in 1984. And received his Master of Science in 1994 in the following subjects: Ecology and Environmental Management; Hydrobiology; and Limnology. From 1991-1995 he was an entrepreneur in nature and landscape consulting and became a teacher in biology and geography in Äänekoski 1994.

Anna VÁRI

Anna Vári, Ph.D. of Economics, is leader of the Division on Risk and Participation in the Institute of Sociology, Hungarian Academy of Sciences. Her main fields of interest include environmental policy, risk analysis, conflict management, public participation and decision support. She has been principal investigator of several Hungarian and international research projects sponsored by various funding organisations including, among others, the US National Science Foundation, the International Institute for Applied Systems Analysis, the Regional Environmental Centre for Central and Eastern Europe, the European Bank for Reconstruction and Development, the PHARE Programme of the European Union, the United Nations Development Programme, and the Global Environmental Fund.

Tero VARJORANTA

Director, Nuclear Waste Management and Materials Regulation, STUK.

Anne VÄÄTÄINEN

Ms. Väätäinen is a Senior Adviser for the Ministry of Trade and Industry in the Energy Department and is responsible for nuclear waste management. She is also a government rapporteur, who prepared and presented the Decision in Principle to the government and will be responsible for the final statement of the Ministry on the EIA Report on the final disposal project.

Juhani VIRA

Master of Science in Engineering from Helsinki University of Technology, received in 1975. In 1981 gained Doctorate of Science at Helsinki University of Technology and became a Research Scientist at the Nuclear Engineering Laboratory of Technical Research Centre of Finland (from 1975-1989).

From 1990 until 1995 served as Section Head for Teollisuuden Voima Oy in the Nuclear Waste Office until becoming Manager of Development for Posiva Oy from 1996-2000. Is currently Research Director for Posiva Oy (since 2001).

Seppo VUORI

Dr. Seppo Vuori is Research Manager at VTT Energy in the Nuclear Energy research field. His educational background is technical physics. His doctoral thesis dealt with the computational assessment of environmental impacts of nuclear power. He is specialised in the assessment of safety of nuclear waste management and final disposal. He has been the co-ordinator of the national public sector's nuclear waste management research programme from 1989 to 2001.

Simon WEBSTER

Simon Webster entered the European Commission as an Administrator in 1992, and spent the first five years as a Euratom nuclear safeguards inspector based in Luxembourg. Since 1997, he has been working in Brussels in the Directorate-General for the Environment, where he has specific responsibilities concerning Community policy in the field of radioactive waste management, both in the EU Member States as well as in the EU candidate countries of Central and Eastern Europe. His background is in Physics with several years' experience in the nuclear industry in the UK, first in design and then commissioning/operation of the new generation of AGRs, before moving to Paris to work at the OECD/NEA. There he was responsible for database activities involving nuclear computer codes and nuclear library data, as well as international code-interpretation exercises.

LIST OF PARTICIPANTS

BELGIUM

BERGMANS, Anne   University of Antwerp
                                     Dept. of Social and Political Sciences

CLAES, Jef   General Manager Belgoprocess

HOOFT, Evelyn   Press Office, ONDRAF/NIRAS

VANHOVE, Valentine   Head, Communication Services
                                     ONDRAF/NIRAS

CANADA

LÉTOURNEAU, Carmel   Natural Resources Canada

CZECH REPUBLIC

WOLLER, Frantisek   Radioactive Waste Repository Authority

FINLAND

ANDERSSON, Janina   Member of Parliament
                                     Member of the Green Parliamentary Group

HOKKANEN, Pekka   Researcher, Nuclear Waste Management Research
                                     Programme

JANTUNEN, Jorma   Senior Adviser, Finnish Environment
                                     Administration

LAAKSONEN, Jukka   General Director, Radiation and Nuclear Safety
                                     Authority of Finland (STUK)

LESKINEN, Antti   Researcher on EIA, Diskurssi Oy

LITMANEN, Tapio	Senior Assistant (Researcher on communications), University of Jyväskylä
LUCANDER, Altti	Member of the Eurajoki Municipality Council
MANNINEN, Jussi	Deputy Director-General (Nuclear Energy) Ministry of Trade and Industry (Energy Department)
ROSENBERG, Thomas	Sociologist and Chairman of the former Lovisa Movement former Citizen's Movement Against Nuclear Waste Disposal in Lovisa
RUOSTETSAARI, Ilkka	Senior Researcher (Academy of Finland) University of Tampere
RYHÄNEN, Veijo	Managing Director, Posiva Oy
SEPPÄLÄ, Timo	Communications manager Posiva Oy
TUIKKA, Kimmo	Kivetty liike (opponent group to the investigation site)
VARJORANTA, Tero	Director, Nuclear Waste and Materials Regulation, STUK
VÄÄTÄINEN, Anne	Senior Adviser, Ministry of Trade and Industry (Energy Department)
VIRA, Juhani	Research Manager, Posiva Oy
VUORI, Seppo	Research Manager, VTT Energy

FRANCE

LE BARS, Yves	President, Andra
MAYS, Claire	Institut Symlog
MERCERON, Thierry	Andra

GERMANY

APPEL, Detlef	Arbeitskreis Auswahlverfahren Endlagersta
LENNARTZ, Hans-Albert	WIBERA Wirtschaftsbertungs AG

HUNGARY

ORMAI, Peter

Public Agency for Radioactive Waste Management

VÁRI, Anna

Hungarian Academy of Sciences
Institute of Sociology

JAPAN

TAKEUCHI, Mitsuo

Group Manager, Safety Affairs, Science an
Technology Division, Nuclear Waste
Management Organisation of Japan (NUMO)

TAKEUCHI, Shunya

Nuclear Waste Management
Organization of Japan (NUMO)

SAKUMA, Hideki

Senior Research Co-ordinator
Nuclear Cycle Backend Division
Japan Nuclear Cycle Development Institute

SPAIN

LANG-LENTON, Jorge

Director of Communications
Empresa Nacional de Residuos Radiactivos

RUIZ LOPEZ, Carmen

Head, High-level Waste Services
Consejo de Seguridad Nuclear

VILA D'ABADAL, Mariano

M & B Advocats
A.M.A.C. – GMF (General Secretary)

SWEDEN

ENGSTRÖM, Saida

Swedish Nuclear Fuel and Waste Management Co.
(SKB)

HEDBERG, Bjorn

Department of Waste Management
and Environmental Protection
Swedish Radiation Protection Institute (SSI)

SODERBERG, Olof

Special Advisor for Nuclear Waste Disposal
Ministry of the Environment

WESTERLIND, Magnus

Director, office of Nuclear Waste Safety
Swedish Nuclear Power Inspectorate (SKI)

SWITZERLAND

AEBERSOLD, Michael   Federal Office of Energy

FRITSCHI, Markus   NAGRA

UNITED KINGDOM

ATHERTON, Elizabeth   UK Nirex Ltd

UNITED STATES OF AMERICA

ISAACS, Thomas H.   Director, Policy, Planning and Special Studies
                                      Lawrence Livermore National Laboratories

LESLIE Bret   HLW Public Outreach Team
                                      US Nuclear Regulatory Commission

KUSAFUKA, Minako   Graduate School of Geography and
                                      George Perkins Marsh Institute
                                      Clark University

MONROE, Scott D.   US EPA Headquarters

INTERNATIONAL ORGANISATIONS

WEBSTER, Simon   European Commission
                                      DG-Energy and Transport

BELL, Michael   International Atomic Energy Agency (IAEA)

LINSLEY, Gordon S.   Head, Waste Safety Section
                                      Division of Radiation and Waste Safety
                                      International Atomic Energy Agency

BOUDER, Frédéric   OECD/PUMA

KESSLER, Carol   Deputy Director-General
                                      OECD/NEA

PESCATORE, Claudio   Radiation Protection and
                                      Waste Management Division
                                      OECD Nuclear Energy Agency

ALSO AVAILABLE

NEA Publications of General Interest

2001 Annual Report (2002)   *Free: paper or web.*

NEA News
ISSN 1605-9581   Yearly subscription: € 37 US$ 45 GBP 26 ¥ 4 800

Geologic Disposal of Radioactive Waste in Perspective (2000)
ISBN 92-64-18425-2   Price: € 20 US$ 20 GBP 12 ¥ 2 050

Radioactive Waste Management

Establishing and Communicating Confidence in the Safety of Deep Geologic Disposal (2002)
ISBN 92-64-09782-1   Price: € 45 US$ 40 £ 28 ¥ 5 150

Radionuclide Retention in Geologic Media (2002)
ISBN 92-64-19695-1   Price: € 55 US$ 49 £ 34 ¥ 5 550

Using Thermodynamic Sorption Models for Guiding Radioelement Distribution Coefficient (KD) Investigations – A Status Report (2001)
ISBN 92-64-18679-4   Price: € 50 US$ 45 £ 31 ¥ 5 050

Gas Generation and Migration in Radioactive Waste Disposal – Safety-relevant Issues (2001)
ISBN 92-64-18672-7   Price: € 45 US$ 39 £ 27 ¥ 4 300

Confidence in Models of Radionuclide Transport for Site-specific Assessment (2001)
ISBN 92-64-18620-4   Price: € 96 US$ 84 £ 58 ¥ 9 100

The Process of Stepwise Decision Making in Finland for the Disposal of Spent Nuclear Fuel (2002)
      In preparation

The Way Forward in Radiological Protection (2002)
ISBN 92-64-18489-9   *Free: paper or web.*

An International Peer Review of the Yucca Mountain Project TSPA-SR (2002)
ISBN 92-64-18477-5   *Free: paper or web.*

GEOTRAP: Radionuclide Migration in Geologic Heterogeneous Media (2002)
ISBN 92-64-18479-1   *Free: paper or web.*

The Role of Underground Laboratories in Nuclear Waste Disposal Programmes (2001)
ISBN 92-64-18472-4   *Free: paper or web.*

Nuclear Waste Bulletin – Update on Waste Management Policies and Programmes, No 14, 2000 Edition (2001)
ISBN 92-64-18461-9   *Free: paper or web.*

Order form on reverse side.

ORDER FORM

OECD Nuclear Energy Agency, 12 boulevard des Iles, F-92130 Issy-les-Moulineaux, France
Tel. 33 (0)1 45 24 10 15, Fax 33 (0)1 45 24 11 10, E-mail: nea@nea.fr, Internet: www.nea.fr

Qty	Title	ISBN	Price	Amount
		Total		

Charge my credit card ❑ VISA ❑ Mastercard ❑ Eurocard ❑ American Express

Card No.	Expiration date	Signature
Name		
Address	Country	
Telephone	Fax	
E-mail		

OECD PUBLICATIONS, 2, rue André-Pascal, 75775 PARIS CEDEX 16
PRINTED IN FRANCE
(66 2002 16 1 P) – No. 52711 2002